# IPTV Security

# IPTV Security

## Protecting High-Value Digital Contents

**David Ramirez**
*Alcatel-Lucent, UK*

John Wiley & Sons, Ltd

*Other Wiley Editorial Offices*

John Wiley & Sons Inc., 111 River Street, Hoboken, NJ 07030, USA

Jossey-Bass, 989 Market Street, San Francisco, CA 94103-1741, USA

Wiley-VCH Verlag GmbH, Boschstr. 12, D-69469 Weinheim, Germany

John Wiley & Sons Australia Ltd, 42 McDougall Street, Milton, Queensland 4064, Australia

John Wiley & Sons (Asia) Pte Ltd, 2 Clementi Loop, #02-01, Jin Xing Distripark, Singapore 129809

John Wiley & Sons Canada Ltd, 6045 Freemont Blvd, Mississauga, ONT, L5R 4J3, Canada

Wiley also publishes its books in a variety of electronic formats. Some content that appears in print may not be
available in electronic books.

*Library of Congress Cataloging in Publication Data*

Ramirez, David.
    IPTV security : protecting high-value digital contents / David Ramirez.
        p.    cm.
    Includes index.
    ISBN 978-0-470-51924-0 (cloth)
    1. Internet television.    2. Computer security.    I. Title.
    TK5105.887.R36 2008
    621.388—dc22

                                        2007039302

*British Library Cataloguing in Publication Data*

A catalogue record for this book is available from the British Library

ISBN 978-0-470-51924-0 (HB)

Typeset in 10/12pt Times by Integra Software Services Pvt. Ltd, Pondicherry, India
Printed and bound in Great Britain by Antony Rowe Ltd, Chippenham, England.
This book is printed on acid-free paper responsibly manufactured from sustainable forestry in which at least two
trees are planted for each one used for paper production.

I would like to take this opportunity to give special thanks to Luis Eduardo Niño for taking a chance and trying my ideas, even if they were based more on hope and ambition than on experience.

Also, I would like to give special thanks to Ramon Alonso Jaramillo for seeing beyond the obvious and allowing me to learn, and to Carlos Mario Toro and John Cuervo who guided my work and shared my enthusiasm for security.

# Contents

# Preface

Paraphrasing the famous quote from Karl Marx, I would say that television is the opium of the masses. If we have any doubts, we just need to look at the number of people glued to the TV every day. I fully understand this inclination. When I was young I spent most of my time looking at the world through the TV. Many images and sounds that now as an adult I try to revisit in person. For many of us, black-and-white TV is still a memory (not just a scary story or an urban myth!). We lived with just a few TV channels that started in the morning and by late afternoon were finished. Only in recent years have we had access to cable packages with hundreds of channels and basically any topic we may want to see.

For many years, TV has been a central mechanism for sharing culture. Although books, music and radio are helpful in bringing an insight into other worlds, audiovisual messages are more powerful and gain more attention from the audience. TV is also cheaper than live performances, and the audience is constantly growing as the number of TV sets per family increases. In many countries, TV channels are closely controlled by the political power, which ensures that only acceptable contents are presented to the public. New technologies may change this environment, allowing subscribers to choose what they see and select from different sources worldwide.

Being a TV fan, it was very interesting to get involved in the topic of IPTV. It was almost by accident that I was requested to write a chapter for an IPTV book in 2005. I had to jump head first into the subject and learn as much as I could about IPTV. One of the conclusions from my initial research on the topic was that information was limited, mostly linked with specific products, and some information lacked structure. This is a common situation with new technologies – there are very clever people developing the technology and they have little time to share all the details with the world.

I expanded the topic of IPTV in my MSc dissertation and, as a result of this additional research, concluded that writing a book on the specific aspects of security could be a positive contribution for those interested in the subject. The writing process became a very interesting journey as I was faced with the challenge of structuring in a coherent way a number of separate areas that span different knowledge domains. I tried to replicate my learning process in the book, bringing together all the diverse subjects that form IPTV in a single document that would allow the reader quickly to gain insight into the components and interactions within IPTV environments.

In general, most of the information available on the subject was either related to particular products or was work in progress expected to become a standard in the future. The book intends to provide detailed information about the different elements that comprise the IPTV environment, filling in some of the gaps left by available information.

The most exciting part of exploring IPTV is realizing how subscribers will have the power to control most aspects of their viewing experience. It may not start with the death of television as we know it, but in years to come subscribers will be able to choose exactly when and what type of content they want to access. Today we have a few IPTV deployments worldwide, and these are slowly gathering momentum. This technology will definitely become an alternative to satellite and cable.

Moreover, as we have seen with many other technologies, the first versions do have security vulnerabilities. More specifically, IPTV is a highly complex environment that brings together technologies from many different vendors, and this increases the potential for security problems. The journey of exploring the security of the IPTV environment clearly shows that there are hundreds of potential points of failure. Many components can become the weakest link and allow intruders to have access to digital assets or components within the IPTV environment.

Hopefully, this book will help security professionals gain a broader picture of the challenges and tools available to secure the environment and ensure that security incidents are reduced and controlled.

# About the Author

**David Ramirez**
Senior Manager
Alcatel-Lucent Services

David Ramirez has been involved with information security for the past 13 years. He began his career as a networking specialist and then joined a consulting company managing information risk management practice where he was involved in risk assessments for more than 80 companies. His next move was to a risk management company in the UK, as part of their new information security division. In that role, Ramirez was responsible for developing the methodologies for the practice, including penetration testing, ISO 17799 compliance and disaster recovery. He was involved in security projects for banks and other financial institutions around the world. The projects focused on security awareness, disaster recovery and business continuity, security policies, security architecture, managed security services and compliance with international standards.

Ramirez is a member of Alcatel-Lucent's security consulting practice. His responsibilities include innovation and technology, thought leadership and knowledge sharing.

# 1

# Introduction to IPTV

## 1.1 Introduction

Television is one of the inventions that has shaped the way society and culture has evolved in the past 60 years. Back in 1940, the first commercial television broadcast started a revolution, showing people of all ages how others lived outside their towns and cities. Television had a powerful effect, shrinking the world and creating a unified view of how things were.

In 1969, ARPANET was created, and a new stage in communications started. Then, in 1983, the core protocol of ARPANET went from NCP (Network Control Protocol) to TCP/IP (Transfer Control Protocol/Internet Protocol) and the Internet was born.

Both the TV and the Internet have revolutionized the way we live. We now have TV channels providing information 24 hours a day, and the Internet facilitating both communication and commerce. Several common areas between the two have finally drawn the technologies into merging, creating IPTV (Internet Protocol Television).

There are some differences between IPTV and IP video. Although the two terms are very similar, there is a clear distinction in the way the market is using the two. IPTV can be used to refer to commercial offerings by service providers with very close access to the subscriber and offers a number of TV channels with a similar look and feel to standard television. IP video is more common within websites and portals, offering downloadable contents and, in some cases, even TV shows and movies downloaded on demand. If it has a number of channels and acceptable quality, it would be called IPTV.

IPTV is a new technology that enables much more flexibility to manage contents and facilitates direct interaction with the sources of content, improving the feedback and future planning. The customer experience is greatly improved by allowing more control over the type of contents immediately available, as well as two-way communication with content providers.

A few years ago, another new technology shocked the entertainment industry – the infamous Napster enabled people to share music and movies in an unprecedented way. With

*IPTV Security: Protecting High-Value Digital Contents*   David Ramirez
© 2008 Alcatel-Lucent. All Rights Reserved

this technology it was not just the case of a neighbor lending a VHS tape with an old movie. With Napster, people shared prerelease albums and videos, creating significant losses for the music industry and movie studios.

Napster was eventually shut down in 2001, but several peer-to-peer (P2P) networks appeared and the phenomenon grew dramatically, reaching millions of users worldwide. Checking e-mule would confirm an average user base of 600–900 million users worldwide.

At the same time, several providers have started to offer legal downloads to the general public. Anyone can buy music and video files. The entertainment industry has added digital rights management (DRM) capabilities to the files and applications used to reproduce the contents, which enables a sustainable model for sales of digital content. Recently, some online stores have even removed DRM to calm the complaints from their subscribers related to fair use of the contents. Users feel that, once they have paid for content, they should be able to enjoy it on any device, and DRM is blocking that fair use possibility.

The recently born IPTV industry will need to address the same issues that once affected the digital media distributors. Customers tend to share information, and over the years there have been a number of very clever pieces of software that enable people to share information and content. A recent example of the phenomenon is Freenet, a reportedly headless network of nodes, storing encrypted sections of content and sending it to anyone who requests a particular piece of data. With Freenet it is very difficult to find who is sharing illegal material, and hence the enforcement of intellectual property rights and copyright restrictions becomes more difficult.

One of the main risks faced by the industry is the rise of thousands of 'home-made stations' willing to broadcast DRM-protected contents. One example of the technology that will come in the future is VideoLAN. This software enables multimedia streaming of MPEG-2, MPEG-4, DVDs, satellite and terrestrial TV on a high-bandwidth network broadcast or unicast. If Freenet and VideoLAN meet, then there will be thousands of encrypted stations broadcasting content outside any control of regulators.

However, the IPTV industry not only has DRM and content protection issues, customers are used to an always-on service with consistent quality. IPTV would have to maintain high levels of availability to convince subscribers that this is a viable option.

With a worldwide trend in privacy protection laws, all the information sent and received from the customer must be protected from third parties trying to capture information. The wireless LAN/WAN markets are a prime example that bad publicity happens to good people. IT managers are not purchasing the technology because of fear, uncertainty and doubt around the potential risks of deploying wireless networks.

Many problems that have affected the cable and satellite industry in the past will gradually migrate to the IPTV service providers, with the increased impact of IPTV providing a two-way communication that includes logical paths connecting TVs to the Internet, and with that environment come computer worms and viruses. IPTV service providers must ensure that subscribers are not able to attack the servers providing contents, and also protect subscribers from the Internet and other subscribers. Most importantly, the shared infrastructure with other services has to be protected.

All those risks and threats must be addressed to achieve a profitable business model. The following chapters of this book will cover some of the basic measures required to implement IPTV security.

Chapter 1 will cover an initial reference to threats to IPTV infrastructures, including known attacks and effects on the IPTV solution.

Chapter 2 will cover references to the IPTV architecture, operation, elements and known requirements. This will provide the novice with background to understand the technology.

Chapter 3, under the title of Intellectual Property, will cover the requirements that content owners have placed on service providers to protect contents from unauthorized access.

Chapter 4 provides a technical overview of the threats faced by IPTV and how these can affect the infrastructure and applications.

Chapter 5 is based on the International Telecommunications Union (ITU) X.805, a standard that covers end-to-end security for communication networks.

Chapter 6 will provide a summary of the technology, threats and countermeasures.

The material found in this book will allow readers to understand the basic concepts supporting IPTV and existing threats to the IPTV environment, and will provide a structured approach to defining what countermeasures are relevant and required for the appropriate protection of the IPTV environment.

## 1.2 General Threats to IPTV Deployments

IPTV market growth and adoption is benefiting from the increased bandwidth available as part of new broadband services on a number of different technologies. DSL, cable, mobile phones and Wimax are just a few examples of the type of technologies now offering enough bandwidth for acceptable service levels and customer experience.

It is important to remember that the IPTV business model is based on the general public being able to access intellectual property owned by third parties and being distributed by service providers. Both content owners and service providers derive their revenues from the secure operation of the service. If content were disclosed in digital form and full quality, then the potential revenue would be greatly reduced. The symbiotic relationship between content owners and service providers depends on the use of technological mechanisms to reduce the risk of unauthorized release of the digital media. Most cases include the use of DRM and other security solutions to ensure control over the distribution and access.

What are the threats, risks and vulnerabilities that the industry is trying to overcome?

There are two main areas of concern:

1. The underlying communication technology used to send the content to the subscribers. This is composed of the networking equipment and communication equipment linking the display to the source of data.
2. The second area is the IPTV-related equipment. This is a series of elements designed to operate the IPTV service and provide access and information to enable the service to operate.

Compared with traditional voice/data networks or cable TV infrastructure, threats to an IPTV environment are far more severe. The whole environment can be affected by a single computer worm. IPTV environments are formed by homogeneous hardware and software platforms. In most cases, one or two operating systems would be used for all the set top boxes deployed, but, if a computer worm were to affect the network, then a minimum of 50% of all set top boxes (and subscribers) would be out of service for a period of time. Carriers also need to ensure that quality of service is protected to comply with customer's expectations and service level agreements (SLA).

Those two main areas of concern can be translated into specific threats and risks to the IPTV service.

## 1.2.1 Access Fraud

Access fraud is one of the oldest forms of fraud within premium/paid TV. This situation happens when an individual circumvents the conventional access mechanisms to gain unauthorized access to TV contents without paying a subscription or increasing the access granted.

An example of the type of threats faced by IPTV vendors comes from the satellite TV industry. For years they have been fighting access fraud. The widespread nature of fraud has caused, during recent years, some satellite TV companies to start taking legal action against defendants for unauthorized access to TV content. A whole industry was developed around the provisioning of modified access cards allowing unlimited access to TV packages and eroding the revenue of satellite TV vendors.

The experience of the satellite TV industry shows that fraudsters go to great lengths to break the existing security measures. This includes cracking the smart card protection used for the set top boxes and distributing cloned 'free access' cards. Even though the satellite TV providers modified the cards, fraudsters have managed to find alternative ways to break the safeguards incorporated in the new releases, and this cycle is repeated constantly.

Now that video technology has entered the IP world, the level of threats has escalated – vulnerabilities that have been solved in other, more mature technologies are still part of the new IPTV systems. There is a recent example of a major TV provider stopping their online content distribution owing to security vulnerabilities being found and exploited on the digital rights management technology protecting the content. There could be numerous vulnerabilities discovered on IPTV systems while the infrastructure reaches a higher maturity level. It is important to ensure that the underlying platform has the state of the art in relation to security mechanisms and procedures. This will add protection layers to the environment and will limit the effect of vulnerabilities discovered.

Another relevant example is the constant battle between cable operators and users. In many cases, cable modems have been modified to uncap the access to the network. This situation is presented when someone has access to the configuration function of the cable modem via the software interface or, in some cases, even access to hardware components within the cable modem and the bandwidth and other restrictions are removed. There are sites on the Internet where modified cable modems are offered, as well as kits and instructions to modify the configuration and remove the bandwidth limitations.

IPTV is transferred not only to set top boxes but also to computers and handheld devices. This facilitates the process of breaking the security of contents. Intruders could manipulate or modify the behavior of their IPTV client and extract the content in digital form ready to be copied or broadcast. Simple software modifications introduced by hackers allow them to break the encryption system and other security measures, or even capture and redistribute the contents using peer-to-peer networks.

The main fact related to access fraud is that, in order for an IPTV system to work, end-users have to be provided with the encrypted content, encryption algorithm and the encryption keys. Anyone familiar with these technologies will tell you that you have lost the game at that point as you no longer have control over the content. Historically, these types of environment show that eventually someone will be able to break the protections and release the content.

Access fraud is reduced greatly by the implementation of different technologies intended to block any attempt at unauthorized access, for example:

- The STB has a DRM client needing to liaise with the DRM application to receive the valid keys for the content. Any third party with access to the content will not be able to decrypt the information as no valid keys have been issued for them.
- Communication with the middleware servers is protected using SSL, and STBs can be authenticated, ensuring that only valid systems are accessing the content.
- DSLAMs are able to validate that only valid subscribers are able to connect to the network and communicate with the middleware servers. The physical line used for access to the network is mapped with the MAC and IP address used by the subscriber and is validated to ensure authorized access. The DSLAM will block any access between systems, avoiding peer-to-peer connections that may result in hacking incidents or unauthorized access to content.

### 1.2.2 Unauthorized Broadcasting

IPTV contents are distributed in digital format, simplifying the work of any individual with an interest in copying or broadcasting the contents. One of the arguments in the campaign against movie piracy is that bootleg DVDs tend to be recordings made at the cinema by people using handheld cameras. However, with digital content broadcast as part of an IPTV service there is no difference between pirate and original content.

A major impact on the satellite TV industry has been fraudsters selling modified 'all access' smart cards based on modifications to valid smart cards and receivers. If fraudsters are successful at the same type of attack within an IPTV environment, they will be able to create 'all access' IPTV set top boxes or cards. As a result, the IPTV industry faces an entirely new threat – with broadcasting stations residing on every home PC, hackers would be able to redistribute the broadcast stream to other computers all over the world. There are some known cases where individuals have offered redistribution of sport events, charging interested people a small fraction of the commercial cost of accessing the content.

Taking as an example the widespread effect of peer-to-peer networks and how easy it is to use one of these environments to distribute large amounts of data, it is technically feasible to set up a peer-to-peer network used to distribute broadcast IPTV content. One single source could be used to deliver contents to nonpaying viewers around the world. This is a clear danger to the business model followed by content owners and service providers. A valid subscriber could take digital content and use peer-to-peer networks to distribute the high-quality content to a large audience, eliminating the need for those viewers to pay for the content or maintain any subscription to commercial TV services. All the technology pieces are available for this situation to arise.

### 1.2.3 Access Interruption

Television is a service that people take for granted – the public expects to click the button and get something on the screen. If an intruder were able to damage the infrastructure or one of the service components, then customers would loose access to their services, causing a loss of confidence in the service. Cable operators offer a pretty much reliable service, and customers would compare the reliability of IPTV networks against other solutions.

Security and reliability must be built into the architecture to ensure that the service is always available and any interruptions are quickly solved.

The way most IPTV solutions are deployed creates a number of risks, especially from fast-replicating attacks such as the ones from worms and viruses. A worm capable of attacking the set top boxes could bring down several hundred thousand boxes in seconds and, properly coded, would cause an outage of weeks while technical support people recovered the boxes to their original state. Similar attacks could be launched against web-based middleware servers, leaving all viewers without access to their electronic programming guide.

STBs tend to have the same operating system within a particular service provider. If the central server were infected by a worm or virus, it would be a matter of seconds before all STBs were infected, easily bringing the service down.

The major weaknesses within the IPTV environment, related to access interruption, are as follows:

- Middleware servers, even if deployed in a high-availability environment, are a single point of failure. If vulnerability were exploited on the servers, then intruders could shut down the middleware servers.
- Denial of service is also a major risk within the middleware servers. If there are no appropriate mechanisms, intruders could send a number of invalid requests to the middleware server, blocking access by valid users.
- DSLAMs tend to have the same operating system. If an intruder is capable of affecting the configuration of a number of DSLAMs, then thousands of users would be left without service. An additional problem is that some DSLAMs tend to be deployed in rural areas with limited access by support personnel, and recovering service may take from several hours to several days.
- STBs tend to run known operating systems, and a worm exploiting vulnerability on those systems could shut down all STBs simultaneously, even disabling the STB permanently until a technician has physical access to the system.
- Residential gateways present the same type of risk. A massive attack could shut down all RGs and leave customers without access.
- There are similar risks within the IPTV core components. For example, if an intruder were to disable the broadcast server or video-on-demand server at the regional head end, thousands of subscribers would loose access to the server. This is valid for the DRM and other IPTV components at the head end. In general, the whole infrastructure should be designed following an approach of high availability.

## 1.2.4 Content Corruption

The resources and funding required to broadcast over-the-air fake signals are so large that this is something usually left for military use. There are no frequent cases of people starting their own TV station and blasting their message to large regions of a city or even across cities. Cable operators have to their advantage that any modification to the signal requires physical access and can be easily tracked.

On IPTV, a different environment is presented as the signal is being sent using normal IP protocols and intruders could connect via the web and manipulate the middleware or broadcast servers. It is also possible to change the data within the content repository before it has been encrypted by the DRM software. An intruder could manipulate a particular movie or content and cause the IPTV provider to broadcast inappropriate or unauthorized content.

Content has to pass through different intermediaries before it is sent to subscribers. There are three main sections of the journey between the content providers and the subscriber:

1. There is an initial path between the content owner (or its agent) and the service provider operating the IPTV service. This can be via satellite, Internet or magnetic media. Any of these can provide an opportunity for unauthorized modification of the content. In some cases encryption is used, but there can be cases where this protection is broken, in particular if there are no appropriate mechanisms to update and manage the keys.
2. Content is then stored by the service provider at the content database, allowing an opportunity for unauthorized access by intruders or employees who could modify the contents. Disgruntled employees could have access to the database and modify the content either by editing or replacing the files.
3. The last stage is the transport between the regional head end and the STB. If there are no appropriate protections, the content could be modified or new content released to the subscribers. Intruders could attempt to insert broadcast traffic to be received by STBs, trying to have STBs displaying the fake content to subscribers.

# 2

# Principles Supporting IPTV

Understanding the underlying principles and mechanisms behind a particular technology facilitates the process of identifying and controlling security vulnerabilities. IPTV cannot be seen as a black box that is formed by a number of products or platforms.

This chapter presents an introduction to the principles supporting IPTV, including the history of moving images and how images are captured and reproduced. Within this section the reader will find references to the physical principles used by the technology, as well as to some of the predecessors of modern components. Scientists and inventors went through a long process before being able to capture and reproduce videos with acceptable quality. Continuous iterations were required before arriving at a viable technology.

The principles behind digital video will be covered, presenting information about how video is codified and what characteristics and parameters are considered to be relevant when dealing with moving images. These principles are critical when analyzing quality of service within IPTV environments, as any disruption will cause a matching degradation of video quality.

References to encoding and compression are presented, as these concepts are one of the key enablers of IPTV by facilitating the transport of vast amounts of video data over relatively small links. Readers will be able to understand the principles behind encoding and compression, as well as some of the standard ratios within IPTV.

The chapter will finish with a very brief reference to TCP/IP for readers coming from a non-IT background. Readers not familiar with TCP/IP would benefit from additional reading available on networking books and generally on the Internet.

## 2.1 History of Video and Television

IPTV is based on video standards for the operation of the service. Camcorders and computers supported the development of a number of technologies and inventions around video, and

---

*IPTV Security: Protecting High-Value Digital Contents*   David Ramirez
© 2008 Alcatel-Lucent. All Rights Reserved

these have been integrated into the IPTV service. The evolution of the technology is paved with clever inventions, each one adding small components that gave form to this new technology.

In order properly to understand the operation and challenges within an IPTV environment, as well as securing the infrastructure against potential threats, it is useful to view the evolution of the technology and how components are interrelated.

This chapter section provides a brief introduction to the history of television, starting with the first technological developments that enabled the reception, transmission and screening of moving images.

## 2.1.1 Television

Television is commonly used to describe the telecommunication system for remotely broadcasting and receiving video and sound. Most people are also familiar with the fact that television can be broadcast via different media, such as over the air and via cable systems, and most recently television is being broadcast using Internet protocol networks (IP networks).

As an invention, television is the result of the constant evolution of scientific knowledge and a number of creative sparks on the part of many of inventors. TV was not an independent invention, and over time it has constantly evolved to include better quality and more flexibility for TV enthusiasts.

- One of the first discoveries supporting television was by Willoughby Smith [1]. This English electrical engineer was working on deploying underwater cables and needed a mechanism to test the cable for ruptures while it was being laid on the sea floor. Smith decided to use selenium bars, which provided a very high current resistance. Coincidentally, Smith found that this resistance changed when the selenium was exposed to light. Smith reported his findings in a letter to the Society of Telegraph Engineers, and also in a subsequent paper published by the same institution. This discovery was made in 1873, and was later used to coat glass surfaces. The characteristics of this material allowed scientists to 'read' images projected onto the coating and subsequently provided the basis of the first TV cameras. TV required not only mechanisms to show images but also ways to capture the image and transmit it over distance. Without a viable way to capture the image, there would be no use for future developments allowing for the projection of images onto glass surfaces.
- Another critical invention that provided support to the development of television was the scanning disc patented by Paul Nipkow. This solution worked by rotating a disc with small holes in spiral form. The holes passed sequentially in front of the image and each one let different intensities of light pass through. The beams of light then reached selenium sensors, thus creating electrical disturbances that could be used to transmit and reproduce an image. By dividing an image into smaller parts scanned by the disc, scientists were able to translate images into electrical impulses. These sensors were still archaic and required a very strong light in order to react. This concept was useful in the first TV demonstrations.
- Most initial television systems used the basic design of scanning an image to produce a time-sequence version of the image. That image representation was obtained in electrical form, and thus it was possible to transmit the signal to a device capable of reversing the

process, creating a reproduction of the image. The reproduction was based on sending an electron beam against a glass wall covered by reacting chemicals that would illuminate when bombarded by the electrons. The initial versions of both the cameras and the screens were very simple, and they were used to show geometric forms or other basic figures. With a number of additional discoveries, this evolved into full human form and then color TV as we currently know it.

- German scientist Johann Heinrich Wilhelm Geissler [2] (1815–1879) invented a glass tube that could be used for demonstrating an electrical discharge. Geissler tubes are a glass recipient containing rarefied gases or conductive liquids as well as an electrode at each end. When high voltages were applied to the terminals, an electrical current was seen flowing through the tube. This phenomenon was caused by the current disassociating electrons from the gas molecules, creating ions, and then the electron flow recombining with ions and causing light to be emitted. Each gas created a different light effect directly related to the chemical composition. During operation, the glass tube would glow owing to the transmission of a ray from the negative cathode at the opposite end of the device. This phenomenon is known as cathode rays. Geissler tubes were very useful for showing how rays traveled and provided a working platform for scientists to find other important applications for the cathode rays. Future developments included bombarding a section of the glass tube that had been coated with reactive material and using the effects of magnetic forces to cause a deflection of the beams. Figure 2.1 shows how the electron beam travels through the Geissler tube.
- English scientist Sir William Crookes delivered a lecture to the British Association for the Advancement of Science in Sheffield (UK) in 1879. He showcased a number of glass tubes based on the principles established by Geissler. These experiments were used to explore the behavior of plasma. One of the tubes used was the 'Maltese Cross', which is a very familiar instrument in modern high-school physics labs. This particular tube was used to demonstrate that electrons had a straight path from source to cathode. This tube comprises a glass cone with an electrically heated wire at the small end of the cone called the cathode, which is the component that produces electrons. On the wide end, a phosphor-coated screen forming an anode is connected to the positive terminal of a voltage source, which in turn is attached to the cathode. A third element is located between the

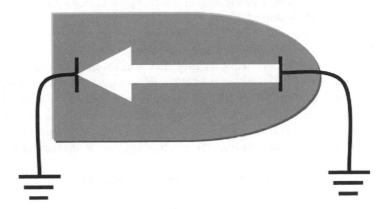

**Figure 2.1**  Example of Geissler tube

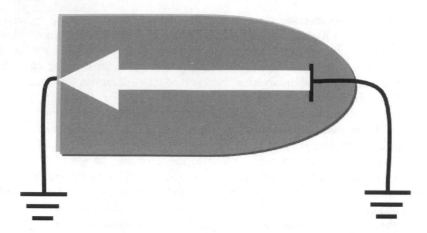

**Figure 2.2**  Example of Crookes tube

anode and cathode. For the Maltese Cross this was a flat, cross-shaped plate which was
also connected to the cathode. Once voltages are applied to the terminals, the phosphor
screen will glow, portraying a shadow of the Maltese Cross. With the demonstration by
Crookes it was clear that electrons formed straight lines when traveling. Other scientists
could explore ways of deflecting the path of electrons to form figures. Figure 2.2 illustrates
the coating at one end of the Crookes tube.

- Different scientists took the Crookes tube and added tests such as tubes with electrostatic
  deflectors (electromagnets) in the horizontal and vertical planes. These enabled scientists
  to observe ray deflections resulting from the specific voltages applied to the deflectors.
  This allowed for experiments showing movement of the beam in response to changing
  voltages. Those changes to the Crookes tube created the basis of the TV screen by showing
  on screen the voltage changes created by the TV cameras. Scientists could deflect the
  beams to reproduce the original images. With these experiments, scientists were able to
  control the beam and the effect on the coated screen. By modulating the voltage changes
  on the deflectors, images could be created on the phosphor-coated surface.
- German Scientist Karl Ferdinand Braun [3] was first to attempt to focus the ray on a
  Crookes tube. Using magnetic fields outside the tube, he was able to create patterns on
  a phosphor-coated screen. This allowed more control over the beam and also allowed for
  the use of electrical magnets to control the rays. This was invented in 1897 and gave birth
  to the cathode ray tube (CRT) concept, the CRT term still used today to describe standard
  TVs and computer monitors. With Braun's invention, the process of creating the basis
  of modern television was finished, scientists were able to capture images in the form of
  electrical impulses, transmit these images over wires and then reconstruct the image using
  the Braun tube. Figure 2.3 depicts how the electron ray can be diverted using magnetic
  fields.
- The work by Braun was expanded by Vladimir Zworykin, who created the iconoscope
  [4], a term used to describe the picture tube inside the television.
- John Logie Baird is identified with the world's first demonstration of a working television
  system based on the Nipkow principle using a rotating disc in 1926. Baird further

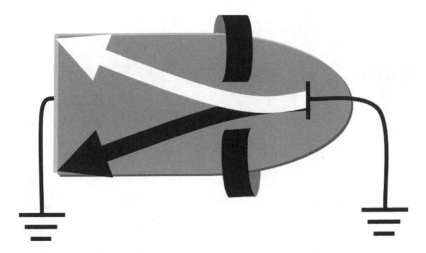

**Figure 2.3**  Example of Braun tube

showcased the world's first color television transmission in 1928. Baird integrated the concepts and developments of his predecessors to support his television system.

- Completely electronic television systems relied on the inventions of Philo Farnsworth, who gave the world's first public demonstration of an all-electronic television system at the Franklin Institute in Philadelphia on 25 August 1934. Philo Farnsworth created the image dissector in 1927. It is understood that his inspiration came while tilling a potato field in Idaho at the age of 14. This particular invention used a Crookes tube to transfer an image into electrical signals. An external image is focused on the surface of the tube, which causes the surface to emit electrons that can be directed into a detector that absorbs the electrons. After sequential scanning of the surface, this allows for a representation of the image. The image dissector provided a more effective way of capturing images. Previous systems using rotating discs had an extremely low definition and required high-intensity images. Figure 2.4 shows an example of a letter T being focused on the surface of the Farnsworth tube. This creates a charge on the surface, following the shape of the original image. Figure 2.5 illustrates how the electron ray can be used to scan the surface and release the charge via one of the terminals of the tube. The resulting electrons will be a linear representation of the image and can be transported and visualized on an inverted tube projecting an electron ray that is diverted by magnetic fields that follow voltage variations.
- German scientist Vladimir Zworykin created the concept of projecting an image on a screen covered with a photoemissive assortment of granules of material, a pattern equivalent to the structure of eyes within the animal kingdom. As each granule has different amounts of light hitting at any given time, a charge image is formed on the surface. A plate behind the mosaic was used to create a temporary capacitor, and sending an electron beam against the plate created changes in potential at the metal plate which represented the picture.
- The science behind both capturing an image and reproducing the image took more than a century to evolve. Today this technology has been translated to silicon, and images can be captured using silicon-based elements, creating a signal that can be easily reproduced

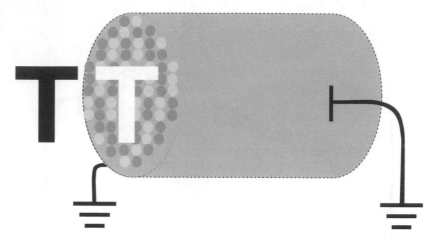

**Figure 2.4**   Farnsworth tube charged with image of letter 'T'

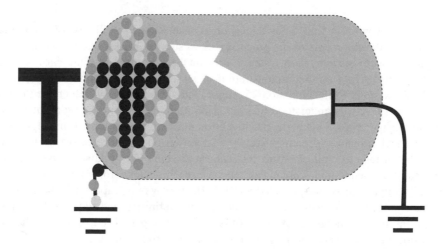

**Figure 2.5**   Farnsworth tube being discharged via collector

on plasma or LCD screens. Video technology was initially used for television systems. Since then it has evolved into many different formats and mechanisms from closed-circuit television (CCTV), personal video recorders, digital video recorders, digital video Disc, high-definition TV and many others, even allowing viewers to use personal computers to have access to videos.

IPTV relies on modern video technology to broadcast contents. Video is captured using high- and standard-definition cameras creating high-quality feeds that, in turn, are either stored or transmitted to the IPTV service provider for distribution among subscribers. Video contents can be found in both digital and analog form, and service providers would have to encode and broadcast the data on the basis of the capabilities of set top boxes and TV sets.

At present, the term 'video' generally refers to numerous storage and reproduction formats for moving images: DVD, QuickTime, MPEG-4, VH, etc. The underlying technology supporting video provides functionalities and flexibility to the type of physical media used to record and transmit the contents, as in either analog or digital form videos can be stored and broadcast using a variety of mechanisms. For IPTV it is important to note that set top boxes are prepared to receive only video in digital form that has been encapsulated for transmission using TCP/IP. Standard analog broadcast is not possible within an IPTV environment, and service providers require mechanisms to encode and encapsulate the feeds.

## 2.2 Viewing Experience of Video

There are a number of aspects that define the viewing experience of video. The human eye and brain have specific ways of perceiving moving images, and with certain refresh frequency it is possible to give the appearance of movement. Some of the key aspects to consider are line scanning, video resolution, number of pictures and aspect radio. All these aspects may be affected by the particular methods of video compression used on the images.

### 2.2.1 Line Scanning

Line scanning refers to the process of sweeping the screen with the electron beam. There are a predetermined number of possible lines on the screen, and the TV set can sweep them in different orders. Within interlacing, the horizontal scan lines of each frame are numbered consecutively and partitioned into two fields: the odd field consisting of the odd-numbered lines and the even field consisting of the even-numbered lines. When refreshing the images, the system will refresh a block of fields (half-frame), which makes it possible to transmit videos over a narrow bandwidth. Each time, the system is sending only half of the image, which reduces the overall load on the transport media. The human eye will perceive a level of change, and with the next image the movement will be completed.

Within the progressive scan systems, each refresh updates all of the lines. Quality and viewer experience are greatly improved. However, bandwidth requirements are much higher.

IPTV requires significant bandwidth availability for appropriate viewer experience. In some cases where the bandwidth is limited (mobile phones or WiMax), interlacing can be used. IPTV subscribers would expect equivalent image quality to terrestrial, satellite and cable TV. Low-bandwidth interlaced scanning may not be acceptable to many viewers.

### 2.2.2 Video Resolution

The most widely used mechanism to measure video resolution is the number of pixels (for digital video) or horizontal scan lines and vertical lines of resolution (for analog video). Pixels are a familiar measure within computer screens, and consumers are used to the reference of width × height on many products. The same format is found on new digital plasma and LCD screens. Standard definition (SD) and high definition (HD) would have different requirements in terms of number of pixels – the higher the number, the better will be the customer experience, but bandwidth will be impacted upon by the definition. On analog broadcast systems it is common to find different equivalent quality levels: the resolution for SD is 400 × 480 pixels (NTSC) and 400 × 576 pixels (PAL) for TV broadcasts.

## 2.2.3 Number of Pictures per Second

The number of pictures per second shown on the screen is usually referred to as frames per second – fps. With an fps of 10 it is possible to create the optical illusion of movement: the higher the fps, the better will be the viewer experience. It is common to find that each international standard for video recommends a different fps, for example the PAL and the SECAM standards specify 25 fps, while NTSC specifies 29.97 fps. The fps value will also have an impact on the bandwidth requirements for IPTV broadcasts, and it is important to select video compression mechanisms that will not take the fps to unacceptable levels.

## 2.2.4 Aspect Ratio

The aspect ratio describes the ratio between the width and height of the screen component. As an example, standard television screens use 4:3. High-definition televisions (new digital screens) use an aspect ratio of 16:9. Videos can be adapted to different aspect ratios, for example a 4:3 video can be shown on 16:9. The aspect ratio is linked with the video resolution. For HD and some new SD contents, the aspect ratio will be 16:9 by default.

Many subjective video quality methods are described in the ITU-T recommendation BT.500.

## 2.2.5 Video Compression Method

When video is manipulated in digital form, it can be compressed to facilitate both storage and transmission with reduced loss of quality. Video data has a number of redundancies, causing inefficiency. There are different compression standards that are discussed in the codecs section and facilitate the compression of video to enable transmission via links with limited bandwidth. IPTV relies on video compression to improve the bandwidth utilization and enable the use of new transmission technologies such as wireless and 3GSM.

## 2.3 Video Compression

Compressor/decompressors (codecs) are a vital technology for digital video applications. These mechanisms support the compression of video content and subsequent reproduction with acceptable quality degradation during the process. This technology is very useful when large amounts of data have to be transferred using limited bandwidth. Codecs can be implemented in hardware or software, depending on specific needs. Each type would have advantages and limitations such as better speed for hardware versions, allowing faster response times and less noise on images, compared with increased flexibility of using software versions of codecs, allowing for updates and changes on the algorithms and code used.

Different codecs have specific functions. Some carry out a translation of the video input from RGB (Red, Green, Blue) format into YCbCr format [5]. YCbCr represents color as brightness and difference signals. Y is the brightness (luma), Cb is blue minus luma (B – Y) and Cr is red minus luma (R – Y). There are advantages of using the YCbCr method. For example, it has better compressibility by providing decorrelation of the color signals. It also splits the luma signal from the chroma signal.

| | | |
|---|---|---|
| <0.384 Mbps | Video Conference | (MPEG-4) |
| 1–2 Mbps | VHS Quality Full Screen | (MPEG-2) |
| 2–3 Mbps | Broadcast NTSC | (MPEG-2) |
| 4–6 Mbps | Broadcast PAL | (MPEG-2) |
| 12–20 Mbps | Broadcast HDTV | (MPEG-2) |
| 27.5–40 Mbps | DVB Satellite Multiplex | (MPEG-2 T.) |
| 32–40 Mbps | Professional HDTV | (MPEG-2) |
| 168 Mbps | Raw NTSC | (Raw) |
| 216 Mbps | Raw PAL | (Raw) |
| 1–1.5 Gbps | Raw HDTV | (Raw) |

**Figure 2.6** Approximate bandwidth requirements – known video types

To visualize the reason why codecs must be used for IPTV-related broadcasts, it is important to know that PAL requires an estimated 216 Mbps of bandwidth, NTSC requires approximately 168 Mbps, while the high-definition TV bandwidth is estimated at 1 Gbps. This makes it almost impossible for DSL and cable modems to support uncompressed TV transmissions and is one of the reasons why codecs have to be deployed. It is important to understand that the codec used to compress the data has to be available at the receiver end. Either by hardware or software implementation, the set top box or computer requires the codec to be installed. This brings additional complexity at the moment of deploying a new codec, as most users will require a hardware refresh before being able to enjoy the new technology.

Figure 2.6 illustrates the different bandwidth requirements within common video types. Raw high-definition television would require a maximum of 1.5 Gbps, while some basic videoconferencing applications would require less than 0.38 Mbps. Compression and codification facilitate the transport via low-bandwidth links, and both the improvement of compression standards and the broad availability of high-speed links are facilitating the development of IPTV.

Some of the commonly used codecs within IPTV are as follows.

## 2.3.1 MPEG-2

MPEG-2 (Moving Picture Experts Group, part 2). Used on DVDs and in most digital video broadcasting and cable distribution systems. It provides support for interlaced video. This codec is being replaced with new versions in spite of the large installed base. Most computer programs use it to visualize DVDs, and internet videos support MPEG-2 as the de facto standard for video.

The MPEG-2 codec is based on the concept that video data will include a high number of redundant sections. By removing the temporal and spatial redundancies, the overall bandwidth required is reduced dramatically. Temporal redundancy is used to describe the characteristic of video data tending to have a similar background on each image. This background remains the same along a number of sequential images, or changes are minimal. Spatial redundancy is a characteristic of video data where some areas of an image are replicated within the same frame of video.

Codecs would have to balance the level of spatial and temporal redundancy within a file. These values would change on different sections of the video. The bit rate requirements of a particular video file would be variable, as different sections could have different compression levels. In some cases buffers are used to achieve a constant bit rate easier to control and transmit, and in some cases the codec would have to drop data in order to comply with bandwidth limitations.

The MPEG-2 codec has been accepted as the international standard by the International Standards Organization. In particular, the Joint Technical Committee 1 (JTC1 on Information Technology) subcommittee 29 (coding of audio, picture, multimedia and hypermedia information) has assigned ISO/IEC 13818 for the MPEG-related standards. At the moment there are 11 entries related to ISO/IEC 13818.

## 2.3.2 H.263

H.263 (ITU-T recommendation H.263). This codec has been published by the International Telecommunications Union under the H Series of recommendations dedicated for audiovisual and multimedia systems. This recommendation covers compression of moving images at low bit rates and is supported by other ITU recommendations including H.261. The low bit rate output allows it to be used for videoconferencing and Internet video. This codec provides an improvement on compression capability for progressive scan video and is widely used on Internet sites for releasing videos.

## 2.3.3 MPEG-4

MPEG-4, part 2 (ISO/IEC 14496). After the success of MPEG-2, the Moving Picture Experts Group developed a new and more flexible standard, intended to bring additional capabilities to video broadcasting and to support the development of digital video. Accepted as an ISO standard in 1999, it has been modified to include a number of extensions. MPEG-4 can be used for Internet video, IPTV broadcast and on storage media, among many other functions. It includes object-oriented coding features, enhancements of compression capability and security mechanisms, and it supports both progressive scan and interlaced video. Over time, new set top boxes and IPTV software applications have been prepared to support this standard, which enables more effective compression and better security for intellectual property rights.

The characteristics embedded on this standard are not intended to replace a digital rights management system. Security elements within MPEG-4 are intended to work as a complement to a number of other security mechanisms within the whole IPTV environment. MPEG-4 data includes syntax and data fields that facilitate the identification of IPR within each file and imply that this information can be used for the decision-making process.

MPEG-4 supports the identification of digital assets by embedding identifying information within data files. This information can be in the form of unique identifiers or key pairs (for example author<</>> Peter Jones). This information can be used by other components of the IPTV service to ensure adherence to IPR defined for a particular content asset.

The optional intellectual property identification (IPI) data fields include information about the contents, type of content and information on rights holders. The MPEG-4 standard includes an open interface that can be used by programmers to connect to the IPI data and use information to make decisions about contents.

## 2.4 TCP/IP Principles

Transfer control protocol/Internet protocol are the standards supporting the transport of IPTV packets from the service provider to the subscriber. These standards have been used to allow the Internet to grow and adapt itself.

### 2.4.1 Addresses

IP uses identifiers to denote members of the network. Every server, workstation, proxy, firewall, router and switch would require an IP address to be able to communicate on the network. The IP address assigned to elements is a 32-bit binary number that, to simplify human interpretation, is represented in four 8-bit octets separated by decimal points.

For example, the 32-bit binary number 11000000101010000111101000010111 can be split into four octets 11000000.10101000.01111010.00010111, with each one of the octets representing a value between 0 and 255:

$$192 = 11000000$$
$$168 = 10101000$$
$$122 = 01111010$$
$$23 = 00010111$$

The final, human-readable address would be 192.168.122.23.

Based on their physical location or functions, network elements can be grouped in logical networks. These logical networks follow similar IP address structures (some of the octets are similar). To extract the information about the network structure, IP uses subnet masks. Subnet masks are also a 32-bit binary number where all the numbers on a particular octet are either 0 or 1. By doing a logical AND operation of the IP address and the subnet mask, it is possible to filter out the network part of the IP address.

One example of a network mast as a 32-bit binary number is 11111111111111111111 111100000000 which can be divided into octets 11111111.11111111.11111111.00000000 and represented in the decimal form 255.255.255.0.

To filter out the network part of the IP address:

- IP address:                 11000000101010000111101000010111
- Subnet filter:              11111111111111111111111100000000
- Logical AND:             11000000101010000111101000000000
- Network address binary:     11000000101010000111101000000000
- Network address octet:      11000000.10101000.01111010.00000000
- Network address decimal:    192.168.122.0

## 2.4.2 Routing

Routing is used to define the process of forwarding IP packets from their source network to a destination network. Routers are the network elements responsible for routing IP packets. They use either static routes or dynamically generated routes based on different algorithms.

Routing tables can also be used as a security mechanism to filter the flow of information from one network to another. These are also one of the principles behind bastion hosts which, with the addition of a number of elements, become what we know now as firewalls.

### 2.4.2.1 IP Packet

Figure 2.7 shows the different fields defined for an IP packet. The 'version' number refers to the IP protocol stack version used. IPv4 is the most widely used, and IPv6 is the new version being deployed worldwide to improve communications and solve some of the shortcomings of IPv4.

The 'header length' represents the size of the header to be processed by the receiving party. The 'total length' represents the total size of the packet (including the header).

'Identification' is a random number added to the packet for recognition of anomalous signatures, used in conjunction with 'flags' and 'fragment offset'.

'Time to live' for the packet will decrease as it passes routers. When exhausted, the packet will be dropped. 'Protocol' represents the type of protocol (ICMP, TCP, UDP, etc).

Header validation is done using the 'header checksum' field.

Figure 2.8 illustrates the simplified structure of TCP packets. The 'source port' and 'destination port' represent the services on the source and destination machine responsible for sending and receiving the packet. For HTTP (web browsing) the destination port would be 80, and for SSL (secure browsing) it would be 443. These two ports are used by the set top box to communicate with the middleware server.

The 'sequence number' is used to understand what position within a stream a particular packet should be assigned to.

The 'acknowledgement number' is used by two parties to confirm that a communication has been started. When the communication starts, the first system will issue a packet with

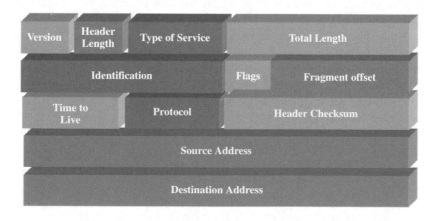

**Figure 2.7**   IP packet – RFC 791

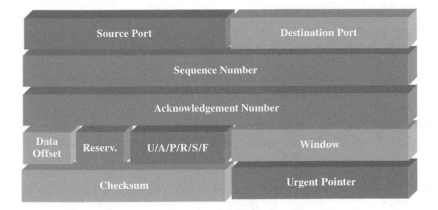

**Figure 2.8** TCP packet

an acknowledgement field value of 0, and on the 'U/A/P/R/S/F' the field of S will be set to 1 and the rest to 0, meaning that this is a SYN packet. The second system will respond with a random value on the acknowledgement field, and on the U/A/P/R/S/F the field of A will be 1 and the rest 0. Finally, the first system will send a packet with the acknowledgement value received from the second system, and on the U/A/P/R/S/F the fields A and S will be 1 and the rest 0.

The U/A/P/R/S/F fields are used for Urgent/Acknowledge/Push/Reset/Sync. Some network intrusion detection systems (NIDSs) tend to use reset flags on packets to block attacks from intruders.

The 'Window' field is used to determine how much data can be sent.

## 2.5 Summary

Great scientists participated in the process of creating the technological basis for recording and reproducing moving images using electronic means. The chance discovery by Willoughby Smith allowed him to translate light changes into current fluctuations, and it also facilitated the development of photosensitive films. The development of the Geissler glass tubes, and all the subsequent variations, allowed scientists and inventors to create displays that could interpret electric fluctuations. Very relevant is the work by Sir William Crookes [6], using the Maltese cross inside the Geissler tube to create a shadow and confirm the linear trajectory of electrons; and the final development by Karl Ferdinand Braun created a direct link between the changes on the magnetic field and the image on the screen. Elements of that type were then improved, signals received enhancements and the concept was used to represent images on screens.

- The ability to capture images and transform them into electrical currents, transport the image and reconstruct it on the screen is one of the key elements of modern television. Current technology uses solid-state circuits instead of coatings to capture images, and plasma or liquid crystal displays are replacing standard TV sets.

- The human eye and brain have specific requirements regarding moving images, and technology has been adapted to provide a superior viewing experience. Some elements must be considered when creating or displaying video:

  - *Video resolution.* This will represent the number of pixels available or used to represent the image: the larger the number, the higher will be the quality.
  - *Pictures per second (pps).* This stands for the number of still images that would be displayed each second: 10 pps is the minimum required for the human eye and brain to have the illusion of movement; higher numbers would improve the quality of the video.
  - *Aspect ratio.* This is a characteristic of the displays and could be very useful in providing the immersion of the viewer into the story. There must be a match between the content aspect ratio and the one supported by the screen. High definition uses an aspect ratio of 16:9.

The original characteristics of the video are critical for the viewing experience and they are sometimes modified owing to technical requirements. Compressor/decompressors (Codecs) are a vital technology for digital video applications. These mechanisms support the compression of video content and subsequent reproduction with acceptable quality degradation during the process. The codec output will have a different size and sometimes will have reduced quality compared with the source. MPEG-2, H.263 and MPEG-4 are some of the most widely used codecs within IPTV deployments.

TCP/IP provides the underlying communication protocol for IPTV, and many countermeasures rely on configuration aspects of TCP/IP. Anyone interested in IPTV security would benefit from understanding how TCP/IP works and the security aspects of filtering and analyzing the traffic.

## References

[1] Lange, A., '*Histoire de la Television*', 2003. Available online: http://histv2.free.fr/selenium/smith.htm [2 October 2007].
[2] Greenslade, T.B., '*Instruments for Natural Philosophy*', 2005. Available online: http://physics.kenyon.edu/EarlyApparatus/Static_Electricity/Geissler_Tubes/Geissler_Tubes.html [2 October 2007].
[3] Braun, F., '*Encyclopdia Britannica*', 2007. Available online: http://www.britannica.com/eb/article-9016270 [6 August 2007].
[4] Schoch, W., 'History of Television', 2001. Available online: http://www2.hs-esslingen.de/telehistory/ikonoskp.html [2 October 2007].
[5] Marjanovic, M., '*What is YlCbCr?*', 2001. Available online: http://www.mir.com/DMG/ycbcr.html [2 October 2007].
[6] 'Oxford Dictionary of National Biography', Oxford University Press, 2007. Available online: http://www.chem.ox.ac.uk/icl/heyes/LanthAct/Biogs/Crookes.html [2 October 2007].

## Bibliography

Bellis, M. (2007), http://inventors.about.com/library/inventors/blnipkov.htm [2 October 2007].
Dijkstra, H. (2007), http://members.chello.nl/~h.dijkstra19/page3.html [2 October 2007].
Jenkins, J.D. (2007), http://www.sparkmuseum.com/GLASS.HTM [2 October 2007].

# 3

# IPTV Architecture

Understanding the IPTV architecture would allow security professionals to form an idea of the threats and countermeasures that would be required. All elements of the IPTV environment are exposed to intruders.

This chapter will start with a high-level view of IPTV, defining a functional architecture and the main components involved in the IPTV ecosystem. The functional architecture is a powerful tool when the responsibility must be shared between different teams. When dealing with a heterogeneous environment, it is important to classify components and explore their interaction and boundaries.

## 3.1 High-level Architecture

The high-level architecture of an IPTV environment comprises four key blocks, each one with particular functions and interdependencies. Within the scope of this chapter are the IPTV service provider, network provider and subscriber [1]. The content provider is only referenced as the source of information, and specific security issues and operation within the content provider are not covered. The main elements of the IPTV environment are depicted in Figure 3.1. Although content owners are outside the scope of this book, they are a critical component of the IPTV ecosystem, and security mechanisms must be deployed to guarantee a safe interaction between IPTV service providers and content owners.

The first block in the figure is formed by the content providers. These can be a number of sources depending on the context. The most typical content providers within commercial/retail IPTV environments are Hollywood studios and entertainment networks. These will sell content packages to IPTV service providers with a variety of contents. In some cases the content will be streamed via satellite links, which allows for a wide geographical distribution of the contents and requires IPTV service providers to set up satellite receivers. Other content providers will issue physical media (DVDs, tapes, etc.) with the content, and

**Figure 3.1** High-level IPTV environment

it will be distributed to a more limited audience. Local content can be delivered via cable or via off-the-air broadcasting. There are some scenarios where the content providers deliver IP ready contents in encoded form. This simplifies the process and reduces the complexity for IPTV service providers.

The second block is formed by the IPTV service providers. These entities are responsible for sourcing content, transforming it into IP content and sending it to subscribers via the network provider. IPTV service providers establish agreements with the content owners, stating whether broadcast content will be encrypted or scrambled to avoid unauthorized access. They will also agree whether video-on-demand and premium content will include digital rights management (DRM) protection to avoid reproduction, replay or storage. IPTV service providers receive a variety of content feeds, and they need to translate those feeds into digital video streams that can be sent via TCP/IP networks. During this transformation, the original content will be modified via encoders to create a lower bandwidth stream. Once contents are ready, IPTV service providers will deliver the content to the subscribers via the network provider. There is a constant interaction between subscribers and IPTV service providers via the network provider. However, this does not imply that both IPTV service providers and network providers should be the same entity.

The network providers are responsible for delivering configuration, status, update and control information from the IPTV service providers to the subscribers, as well as delivering the content requested by subscribers. One of the advantages of IPTV is that any IP network can be used for the function of network provider as long as the bandwidth requirements of the content are fulfilled. This can be seen on mobile networks where subscribers receive contents on their mobile phones using sometimes limited bandwidth capacity. Another added flexibility is that IPTV service providers can use a number of different network providers to deliver content to subscribers. Different networks can be used to deliver content based on the market conditions and expectations from subscribers.

Subscribers are the last element of the infrastructure, they have special equipment configured to receive, interpret and display the contents sent by the IPTV service providers, and they are bound to the license terms agreed with the IPTV service providers. Subscribers are the main asset for an IPTV business, and mechanisms should be deployed to protect subscribers from external actors and even from themselves, as in many cases subscribers tend to follow insecure practices within IP networks.

There are many different combinations for these four blocks. IPTV is not only used within retail environments, and some institutions may decide to create their own IPTV service.

Academic institutions in some countries have decided to deploy their IPTV service and act as content providers, IPTV service providers and network providers, delivering content to students. Corporations and the military may decide to deploy an IPTV service and cover the first three blocks. In each scenario, security experts should consider the specific risks present in a particular environment. This book will consider the four blocks, and specific application of the recommendations of the book will provide insight into the particular risks and threats found in the environment. In retail environments, most IPTV services will be provided by an IPTV service provider acting also as network provider.

### 3.1.1 Service Types

IPTV service providers can offer a number of different services using their infrastructure and capabilities. Some of the most commonly offered services include content broadcast, video on demand (VOD) and games.

Content broadcast is used for normal TV programs and series, using pre-established time slots for particular content and issuing an electronic program guide (EPG) to allow subscribers to watch or record particular contents. Content broadcast uses the multicast capabilities of IP networks, allowing set top boxes to subscribe to multicast domains and receive contents.

Video on demand and pay per view are used for premium content that is sent directly by the IPTV SP to individual subscribers. This requires higher resources and bandwidth and relies on unicast communications.

Both broadcast and VOD services require some specific technologies to operate. IPTV SPs can select to offer both or any of the two as part of their service.

## 3.2 Functional Architecture for the IPTV Service

An IPTV environment can be divided into basic elements. This provides a functional architecture view that allows segregation of duties and specialization. Additional filters can be deployed within these functions, allowing for a more granular view of security. Figure 3.2 shows the six main elements of the functional architecture.

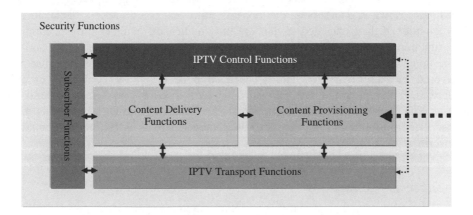

**Figure 3.2**  Functional architecture for IPTV service

The functional architecture is formed by six main blocks representing the following functions: content provisioning, content delivery, IPTV control, IPTV transport, subscriber and security. All are interrelated with security covering all other functions based on their individual requirements and options.

### 3.2.1 Content Provision

All contents used by the IPTV service, including VOD and broadcast material, will pass through the content provisioning function where the encoding (ingest, transcoder, encoder) functions will prepare digital video streams capable of being distributed via the IP network. Within this function, content will also be modified to include commercial or public interest advertisement, logos and brand information as well as security elements such as watermarking and DRM encryption elements.

### 3.2.2 Content Delivery

The content delivery block includes the functions responsible for delivering the encoded streams to subscribers. Information will be obtained from the IPTV control and IPTV transport functions in order to deliver the content to the appropriate subscriber. Content delivery functions will include storage of local copies of contents to expedite delivery, as well as temporary caches stored owing to VOD and network-based personal video recorders (digital video recorders). When subscriber functions interact with IPTV control functions to request specific contents, they will be redirected to the content delivery function to obtain access to the stream.

### 3.2.3 IPTV Control

The IPTV control functions are the hearth of the service. They are responsible for linking together all other functions and ensure that the service is operational at the appropriate levels to guarantee customer satisfaction. From a security point of view, the IPTV control function is important because it acts as the gateway for subscribing requesting contents and it coordinates flows of data among components. Intruders could cause a denial of service (DOS) situation if they managed to damage the middleware or other critical components of this function. The IPTV control function receives requests from subscribers and liaises with the content delivery function and the IPTV transport function to ensure that content is delivered to subscribers. An additional function of the IPTV control is to provide an electronic program guide that can be used by subscribers to select contents to be delivered. The controls will also be responsible for managing the DRM elements required by subscribers to be able to have access to the contents. This is agreed on the contract with the subscriber.

### 3.2.4 Subscriber Functions

The subscriber functions include different actions and elements that are used by subscribers to have access to IPTV contents. Some of these elements are responsible for communication with the transport functions, for example the access gateway that connects to the DSLAM and the rest of the transfer function. Another example is a web browser used by the set top box to connect with the middleware server. Within this function, the set top box stores a number of critical elements including the DRM keys and user authentication information. The subscriber block will use the EPG to allow customers to select which contact to access and request it from the IPTC control functions. It will also receive the digital licenses and DRM keys to access contents. To create the expected output, contents are decoded and displayed.

### 3.2.5 Security

All the functions within the IPTV model are supported by security mechanisms at different levels. Content provisioning will include basic encryption provided by content owners. Content delivery will be ensured using DRM. The IPTV control and transport functions will rely on standards with embedded security to avoid unauthorized modification or access to contents. The subscriber functions will be limited using the security mechanisms deployed at the set top box and middleware server. In general, all applications and operating systems within the IPTV environment should have security mechanisms available to avoid security incidents.

Specific elements from the IPTV environment can be matched with some of the functions. For example, the IPTV control function includes middleware and digital rights management components. When distributing responsibilities, a team in charge of the IPTV control functions will be able to map all applications and components to their function. An example of the mapping can be seen in Figure 3.3.

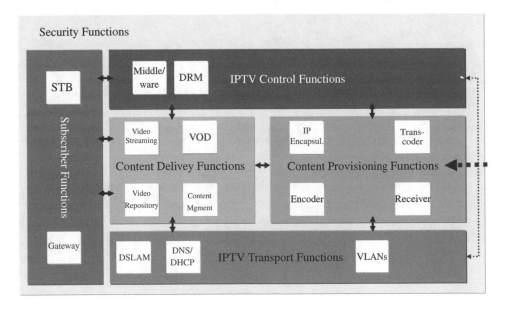

**Figure 3.3**  Components of functional architecture

## 3.3 Detailed IPTV Architecture

The IPTV architecture involves several elements that must be protected individually and as a system to maintain the required security levels.

There are three main sections for an IPTV infrastructure:

1. The head end, comprising all the content feeds, including third-party sources and proprietary contents. The head end falls under complete control of the service provider, and the security and access controls are relatively easy to implement. This may include regional head ends used to support the operation of the service.
2. Home end, comprising the modem and reception equipment including PCs and set top boxes (STBs) where content will be displayed. This is the area where DRM content is decrypted and used.
3. Aggregation network – all the communications equipment used to link the home and head ends.

The three main elements of the IPTV architecture can be seen in Figure 3.4, along with a representation of the main components for the broadcast and video-on-demand functions.

### 3.3.1 Head End (IPTV Service Provider)

The central element of an IPTV infrastructure is the head end. This is formed by a number of elements that receive content, transform it and redistribute it to subscribers following the business model and subscriber packages available. The head end can be deployed using a central head end and regional head ends, which facilitates the broadcast of contents as the regional head ends are closer to subscribers and latency is reduced.

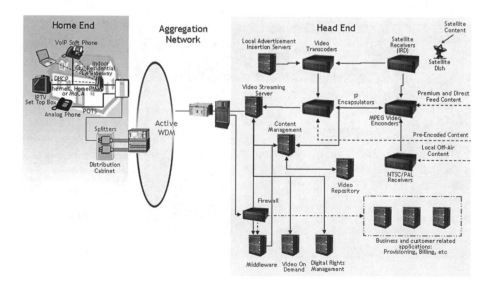

**Figure 3.4** IPTV security architecture

The head end acts as the central point of the infrastructure. It receives all requests from subscribers and provides content to set top boxes accordingly. Additionally, all business-related applications used for provisioning, billing and customer administration are held or linked to the head end. From a security point of view the head end stores the crown jewels, and all efforts must be taken to ensure that access to the head end is controlled, communications within the head end are authorized and content entering the environment is protected at all times.

### 3.3.1.1 Critical Elements of the Head End

*(i) Satellite Receivers*
Integrated receiver decoder (IRD).

*(ii) Video Repository*
This includes:

- video library;
- media library;
- library servers;
- storage area network;
- video-on-demand movie database;
- film server (video and audio files).

*(iii) Content Management System*
This includes:

- command center;
- asset management system;
- digital rights management.

*(iv) Master Video Streaming/Game Server*
This includes:

- propagation service;
- streaming service.

*(v) Ingest Gateway (Video Capture)*
This includes:

- recording system;
- recording manager;
- capture/distribution server.

*(vi) Video Cache Streaming Server*
This includes:

- caching server;
- media cluster.

*(vii) Middleware*
Middleware servers.

*(viii) Business-related Systems*
- accounting;
- provisioning;
- customer information.

The head end receives a series of data feeds in different formats and media, including live feeds from local studios, premium contents from third parties, retransmissions, satellite contents and local video information. Owing to the variety of sources, some of this content is in analog mode and as such cannot be transmitted to the IP network. Contents can be received in DVDs, tapes or video feeds via satellite or over-the-air communications. Contents are then encoded and encapsulated to allow TCP/IP-based transmission. Additional elements involved are the DRM application and content management systems. All communications with the subscribers are coordinated by the middleware server receiving the requests from the different set top boxes.

Figure 3.5 shows the service components of the IPTV environment.

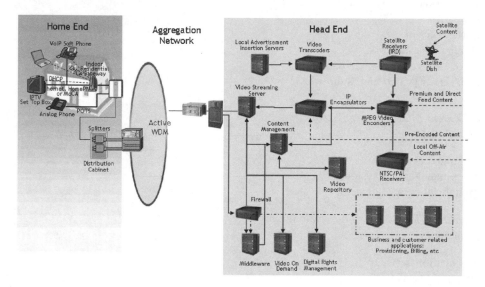

**Figure 3.5**  IPTV security architecture – service components

### 3.3.1.2 Content Input

The following are the feeds and contents received by service providers at the head end:

- satellite content (can be both in analog and in digital form, video or encoded material can be received);
- premium and direct-feed content;
- pre-encoded content ready to be encapsulated;
- local off-the-air content (in NTSC, PAL or ATSC format);
- physical media;
- local advertisement insertion servers.

*(i) Satellite Content*

This requires the use of satellite receivers in the form of an integrated receiver decoder (IRD). Content is broadcast by fixed service satellites with relatively low power and requiring large satellite dishes for reception. The head end receives the radio signal and uses the IRD to tune and amplify the signal. Once the signal is amplified, the IRD can proceed to decode the signal. The IRD can receive both analog and digital signals. All analog input will be sent to the ingest gateway (MPEG video encoder), and the digital signal can be sent to a transcoder to be merged with local advertisement information.

*(ii) Premium and Direct-feed Content*

This tends to be pay-per-view premium contents in analog form and requires encoding by the MPEG encoder before being sent to the IP encapsulator function. The physical media are also classified under this category and are considered direct feed.

*(iii) Pre-encoded Content*

This is content already encoded in an acceptable codec, supported by the set top boxes and software clients. This content feed can be sent directly to the IP encapsulator function.

*(iv) Local Off-the-air Content*

This is content broadcast by local stations and has to be sent to the PAL/NTSC receivers before it is possible to use it.

The different feeds need to be prepared to be used by the IPTV environment. In some cases the analog feed must be transformed into digital/encoded content capable of being encapsulated.

Feeds are processed by the following elements:

- MPEG video encoder;
- IP encapsulator;
- video transcoder.

The relative value and exposure level of the content input are presented in Table 3.1.

**Table 3.1**  Content input asset information

| Asset – Content Input | |
| --- | --- |
| Relative Value | Exposure Level |
| Contents are one of the highest-value assets on the IPTV infrastructure. Apart from customer information (credit card, address and personal information), content is an asset that can be monetized by intruders.<br><br>Most of the content received will be in raw format, making it difficult and impractical to steal. Most intruders will wait until the content | In most environments the exposure level for physical containers will be medium. Physical containers will transport unencrypted contents liable to be stolen or misplaced.<br><br>Satellite, off-the-air and direct feeds tend to be less exposed. This is due either to encryption on the satellite channel or to the characteristics of the unencoded content not being able to travel |

*(continued overleaf)*

**Table 3.1**  (*continued*)

| Asset – Content Input | |
| --- | --- |
| Relative Value | Exposure Level |
| has been encoded with MPEG-2 or MPEG-4, making it easier to transport and share. | through IP networks. Stealing unencoded contents from off-the-air receivers would require physical access.<br><br>In general, content input is exposed to physical access by unauthorized parties. Network security tends to be irrelevant at this point. |

### 3.3.1.3 MPEG Video Encoder

The MPEG video encoder (ingest gateway) is in charge of providing the recording system, recording manager and capture/distribution server used to capture the data and create properly formatted content.

To obtain an appropriate feed, the analog signal is passed through the MPEG video encoders which generate content in digital form ready to be modified and broadcast using the rest of the IPTV components. The process is undertaken by a 'compressor/decompressor' (codec). The easiest reference to this process is a modem taking data and modulating for transmission. The output from the codec is called the 'essence', and it would require additional metadata to enable transmission. Metadata information would include headers, encryption data and other data required to ensure proper transmission. These will be added by additional components down the IPTV chain.

The algorithms used to implement the codec will determine the balance between the quality of the video, the quantity of data needed to represent it (bit rate), robustness to data losses and errors, bandwidth requirements, buffers used and other characteristics. IPTV service providers will determine the most suitable codecs on the basis of the environment, available bandwidth and hardware characteristics of the set top boxes deployed.

Most codecs use a conversion from the standard RGB color format to YCbCr. This conversion improves the compressibility and separates the luma signal. Then the input data will be segmented in blocks separating the chroma and luma data. Some prediction algorithms are used, based on the last image to predict future video data.

For the decoding process, a reverse approach is used, reversing the steps to recreate the video data. During transmission, some information will be lost, and the algorithms and codec must be able to recover from small data losses without affecting the customer experience.

Some of the standard codecs that would be used by IPTV service providers are as follows:

- MPEG-2 [2]. This codec has also being issued by the ITU with the name of H.262. The MPEG-2 is used for digital video broadcasting, cable distribution systems and also for digital video disc (DVD) encoding.
- MPEG-4, part 10. Common areas with H.264. MPEG-4 has dramatic compression enhancements compared with its predecessors.

The output of the MPEG video encoder is digital video ready to be encapsulated, encrypted or stored.

The relative value and exposure level for the MPEG encoder are presented in Table 3.2.

**Table 3.2** MPEG encoder asset information

| Asset – MPEG Encoder | |
| --- | --- |
| Relative Value | Exposure Level |
| This element is responsible for encoding the contents. As such, it would have a valuable output in the form of encoded files that could be easily transported. Intruders having access to these files would be able to distribute the contents and benefit financially from the sale of high-quality digital contents.<br><br>The value of the asset and the output of the process are high. | This is one of the first assets connected to the network and holding the content assets in encoded form. This increases the exposure and threats to the server. Intruders interested in reselling the content in DVD form would prefer contents encoded before they have been encapsulated or processed by the DRM server. |

### 3.3.1.4 IP Encapsulator

The function of the IP encapsulator is to take an encoded video transport stream (TS) and prepare the stream to be broadcast over an IP network. It allows for an encoded video stream to be encapsulated into IP packets that can be carried by the network. Some standards support the transmission of video. One example is real-time transport protocol (RTP). RTP was created to support transmissions of real-time applications such as audio and video. It does not guarantee quality of service or resource allocation, but it includes some functions such as timing reconstruction, loss detection, security and content identification.

The TS input in the form of MPEG or H.264 is translated into IP-based packets that can be sent over Ethernet connections.

The IP encapsulator receives traffic from three main sources: the pre-encoded content being sent directly, the encoded content from the MPEG video encoders and the output from the video transcoder. The output from the IP encapsulator is sent to the video streaming server. Packages can be broadcast or sent to the DRM or content database. In some environments, content reaching the IP encapsulator will be encrypted by the DRM, and some solutions may store the output of the IP encapsulator in the video repository.

From a security point of view, the IP encapsulator creates a highly desirable output as this is ready to be broadcast, as opposed to an MPEG stream/file. If intruders are able to capture traffic or redirect the output, this will be the first point in the chain where they will try to gain access to data.

The relative value and exposure level for the IP encapsulator are presented in Table 3.3.

**Table 3.3** IP encapsulator asset information

| Asset – IP Encapsulator | |
| --- | --- |
| Relative Value | Exposure Level |
| Contents are one of the highest-value assets on the IPTV infrastructure. The output of this system comprises MPEG-2 or MPEG-4 contents | This component creates encapsulated MPEG-2 or MPEG-4 contents. Intruders would be able to connect to the server and extract the |

*(continued overleaf)*

**Table 3.3**   (*continued*)

| Asset – IP Encapsulator | |
| --- | --- |
| Relative Value | Exposure Level |
| encapsulated for distribution via an IP network. It would be relatively easy for an intruder to take the contents and distribute them using a video streaming server or peer-to-peer network. This ease of use would put a high value on the encapsulated contents. | encapsulated files. The only barriers would be the operating system and application controls. |

### 3.3.1.5 Video Transcoder

The video transcoder receives data from the satellite receivers and also data from the local add insertion servers. The main function of the video transcoder is to translate between different codecs. This server will be able to transform a feed from a number of predetermined codecs and generate output on MPEG-2, MPEG-4 or the particular codec chosen by the IPTV service provider. The transcoding process tends to take an encoded stream and convert it into a different encoded stream with a lower bit rate or format without losing too much quality.

With this component, different streams from the satellite or local advertisement can be sent to the IP encapsulator encoded in the preferred standard. A number of algorithms are used to reconstruct the original feed, remove some data without spending too much processing power in the process and generate a stream with comparable video quality encoded with a different standard.

The relative value and exposure level for the video transcoder are presented in Table 3.4.

**Table 3.4**   Video transcoder asset information

| Asset – Video Transcoder | |
| --- | --- |
| Relative Value | Exposure Level |
| Contents are one of the highest-value assets on the IPTV infrastructure. This asset creates encoded contents that could be copied or distributed. The value of these assets is high owing to the possibility of distributing the contents and creating DVD copies. | The exposure of this element is high as intruders would be interested in this particular content once the satellite protections have been lifted. This tends to be premium content that can be easily sold or distributed. |

### 3.3.1.6 Content Management Server

The content management server will control the flow of information from the IP encapsulator and video streaming server, storing all the relevant media into the video repository or sending data to the DRM system. Requests from the middleware server can be serviced by the content management server. This will request data or instruct servers to release it via the video streaming server.

The relative value and exposure level for the content management server are presented in Table 3.5.

**Table 3.5**  Content management server asset information

Asset – Content Management Server

| Relative Value | Exposure Level |
| --- | --- |
| This asset does not hold specific contents and as such the value is low compared with other asset-holding files. This asset is important to ensure the continuity of the operations and also to avoid damage to the reputation and service. Intruders could manipulate the application to replace, rename or modify contents. It is important to protect this asset, but intruders would be more interested in other elements of the IPTV infrastructure. | The exposure level of this asset is lower than that of assets holding contents. Intruders are more determined to obtain digital contents than disturbing the operation. Security measures must be deployed to protect this asset, but this is not one of the prime targets for intruders on the IPTV head end. |

### 3.3.1.7 Video Repository

The video repository is used to store contents in preparation for other applications broadcasting the contents. The storage facility includes a video library and media library hosted on the library servers, ensuring fast and reliable access to required contents. The video repository is a critical element that stores all the digital assets at the disposal of the IPTV operator. From a security point of view, the video repository is one of the most critical elements of the IPTV environments. Most of the digital assets will be stored here, and intruders will try to access this server to extract contents. Content tends to be stored without additional protections including database encryption or DRM protection. Some DRM solutions would support the encryption of contents while stored, and then keys could be distributed to subscribers. Other versions of DRM will not support this option, and contents would be encrypted prior to broadcasting them.

There are also storage facilities for the video-on-demand movie database and in some cases for pay-per-view events.

A phenomenon that occurs with the video repository and DRM systems is that most subscribers will want to access a very small number of titles, whereas a small number of subscribers will request a very large number of titles. This creates a long tail and causes additional storage costs and is necessary to provide acceptable service/content levels to subscribers. Within DRM, service providers could deploy more expensive and complex DRM systems for the highly desirable content and use less expensive DRM mechanisms for the long-tail contents.

The relative value and exposure level for the video repository are presented in Table 3.6.

**Table 3.6**  Video repository asset information

Asset – Video Repository

| Relative Value | Exposure Level |
| --- | --- |
| This server holds the crown jewels. This is the most valuable component, similar | This is the main objective for any intruder on the IPTV network apart from the financial systems. Intruders will target this over any other system as they can gain immediate benefits by downloading |

*(continued overleaf)*

**Table 3.6**  (*continued*)

| Asset – Video Repository | |
|---|---|
| Relative Value | Exposure Level |
| only to the financial systems used by the IPTV service provider.    This server hosts all the digital content, already encoded and ready to be used. | content files as opposed to manipulating financial data that may be detected or traced to them. This remains one of the most exposed servers owing to the willingness of intruders to break into this system.    The architecture of the IPTV head end allows for some basic protections, for example restricting access to the video repository and allowing only a number of approved IP addresses to communicate. |

### 3.3.1.8 Digital Rights Management

The DRM software provides encryption mechanisms that avoid unauthorized access to digital assets and also deploy digital licenses that ensure compliance with the business model defined by the content owner and the IPTV service provider. The DRM server will receive contents after they have been encoded, and in some cases after they have been encapsulated. The payload is then encrypted to avoid unauthorized access to contents. Subscribers are provided with DRM keys that enable them access to the contents under the particular conditions defined by the content owners and the IPTV service providers.

Digital rights management is used to ensure that content is accessed only by authorized subscribers, always following their service terms and conditions. This is encryption technology applied to deliver the business-related rules around the content and ensure that intellectual property rights are respected at all times. The DRM system will encrypt contents to avoid interception between the head end and the home end. Additionally, some contents will have specific restrictions based on the business rules, for example a movie may not be recorded or reproduced. The use of DRM is dependent on the agreement between the content owner and the IPTV operator. In some cases a very light approach to DRM would be used, encrypting only VOD content and leaving the broadcast traffic in clear text. This topic will be discussed in detail in Chapter 4 dealing with intellectual property rights.

The relative value and exposure level for the DRM server are presented in Table 3.7.

**Table 3.7**  DRM asset information

| Asset – Digital Rights Management | |
|---|---|
| Relative Value | Exposure Level |
| This system holds the access keys to the contents distributed to subscribers. As such it has a high value to intruders. If intruders are able to take control of this system, they could create ghost accounts or have access to protected contents and sell the access. | Many intruders will target this system owing to the challenges that it represents and the value of the keys hosted.    Additionally, this asset is exposed to the subscriber network, allowing some interaction for key exchange (in other cases the key exchange will be done via the middleware).    Intruders will target this system over more operational machines. |

### 3.3.1.9  Video Streaming Server

The video streaming server follows the commands from the middleware and video-on-demand servers, and it also receives input from the DRM and content management server with streams in MPEG-4, MPEG-4, H.264 or similar codecs chosen by the IPTV service provider.

The IPTV service provider will need one server to cover a number of users, and they will need to add servers as more subscribers join the network. The video streaming server provides encoded streams that are delivered by the TCP/IP network to the set top boxes. These components tend to support TCP and UDP and multicast streaming depending on the type of application and solution deployed by the IPTV service provider. Different servers will have particular streaming capacity. Depending on the processing power, some servers can handle several thousand subscribers at the same time.

Some basic authentication mechanisms can be deployed within the streaming server, and it would also interact with DRM servers in those cases where the content is not already encrypted ready for distribution.

TCP streaming will present some advantages in relation to flexibility and scope, but it poses an additional load on the streaming server, comparable with other methods. For UPD there are improvements in the number of subscribers supported and the load to the server, but there are possibilities of quality being affected by packet losses.

IPTV service providers can also distribute contents in multicast mode. This type of distribution has a much higher number of subscribers with low-bandwidth utilization compared with unicast mode for TCP and UDP. Each packet is sent only once in a broadcast mode to a multicast VLAN. Set top boxes will join specific multicast VLANs to have access to contents. This avoids any need for individual links between set top boxes and the streaming server.

The relative value and exposure level for the video streaming server are presented in Table 3.8.

**Table 3.8**  Video streaming server asset information

| Asset – Video Streaming Server | |
| --- | --- |
| Relative Value | Exposure Level |
| This system distributes content to subscribers. It does stream valuable content, but in less quantity than stored on the video repository. Intruders may try to redirect or steal the information temporarily stored on this server. The relative value of this asset is medium compared with other, more critical elements. | This particular server is connected directly to subscribers as it needs to support some basic networking exchanges with set top boxes to maintain the flow of packets. If the VOD functions have also been deployed using this server, then some of the packets related to VCR-like controls will have to be allowed to flow from set top boxes to the video streaming server, creating a higher exposure. |

### 3.3.1.10  Subscriber Interaction

The middleware servers are the front end of the IPTV environment. All STBs communicate with the middleware server to request the specific content they require. This communication

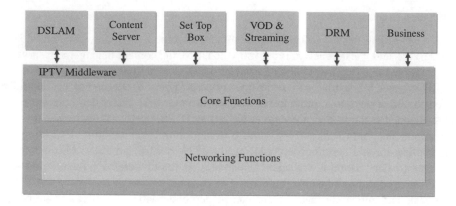

**Figure 3.6**  IPTV middleware architecture

is usually done using HTTP. A browser within the set top box will communicate with the middleware server, downloading the electronic program guide and sending requests to the middleware servers. The middleware server in turn will instruct the streaming server to send contents to the destination. Subscribers are able to communicate via the set top box and request contents from the middleware. Once the middleware has instructed the VOD servers to provide contents to a particular subscriber, then the set top box and VOD will also be able to communicate in order for the network-based personal video recorder (PVR) to work.

There are some IPTV service providers that have enabled web servers on the DMZ to host applications that allow control over the nPVR. These are another case of subscriber interaction and represent an additional risk owing to the permanent connection to the Internet.

The IPTV middleware acts as a broker between a number of systems and applications. More specifically, it interacts with the digital subscriber line access multiplexer (DSLAM), content servers, set top boxes, video-on-demand and content streaming and DRM servers as well as business applications among other systems.

Figure 3.6 illustrates the middleware architecture and how the component interacts with the other core elements of the IPTV environment.

IPTV middleware systems tend to have two main components: a set of core functions and a number of networking functions relying on a standard operating system and providing a web-based interface for set top boxes to communicate. The core functions include management, control of transactions and existing sessions with set top boxes, as well as user authentication and other critical functions. The core functions are responsible for maintaining the EPG and other basic functions as well as coordinating activities of external systems such as content management, DRM, VOD and business applications.

Table 3.9 shows the relative value and exposure level of the middleware server.

Some external systems interacting with the Middleware system are as follows:

*(i) DSLAM*
The digital subscriber line access multiplexer. Some models of DSLAM allow the middleware server to share authentication data with the DSLAM, which facilitates the process of authorizing the access of new subscribers to content VLANS as well as exchanging information about the physical location of set top boxes used by subscribers. The middleware

**Table 3.9** Middleware server asset information

| Asset – Middleware Server | |
| --- | --- |
| Relative Value | Exposure Level |
| The middleware system does not store critical assets, but from there intruders may be able to manipulate how video is provided by subscribers. The relative value in terms of brand integrity is high, as any modification to the electronic program guide will be sent to all subscribers. Additionally, intruders could cause a denial of service against the IPTV infrastructure by knocking down the middleware servers. | This is the most exposed server within the IPTV environment. The middleware server allows web access to the IPTV application, and this facilitates the use of known web service and web application attacks to either take control or shut down the service. It is difficult to filter access to this server, as all set top boxes should be allowed to browse information on the middleware server. The level of exposure is high. |

provides set top boxes with information about the existing VLANs that they must join in order to access contents. From a security point of view, the middleware system could instruct the DSLAM to shut down a physical link if one particular set top box has been identified as a risk to the IPTV environment, a risk to other users or is in breach of the subscriber agreement.

*(ii) Content Server*
The middleware server will receive information from the content server regarding the available contents and will use this information to prepare the EPG.

*(iii) Set Top Box*
There are many different levels of interaction between the middleware server and the set top boxes. Set top boxes are configured to check a particular VLAN each time they start their operating system. This VLAN will provide critical updates required for operation. Once the middleware server has been visited, the set top box could be instructed to download required updates. Most set top boxes will use a web browser to download the EPG and basic information from the middleware server. DRM keys and other critical data may be provided directly by the middleware server, or set top boxes could be directed to the DRM application to obtain keys.

*(iv) Video on Demand and Streaming*
The middleware will receive requests from the set top boxes on specific EPG entries including the VOD and pay-per-view contents. There will be an interaction between the middleware and VOD to start streaming the unicast contents to the subscriber.

*(v) Digital Rights Management*
The middleware application will retrieve keys and digital licenses to be used by set top boxes, in some cases set top boxes will be referred to request the data directly from the DRM server.

*(vi) Business Applications*

The middleware server will interact with business applications for user validation and confirmation, billing functions and settlement, as well as account information requested by subscribers.

## 3.3.2 Transport and Aggregation Network (IPTV Network Provider)

Contents leaving the head end will take one of two forms: it will either be unicast traffic, directed to specific subscribers, or it will be multicast traffic, directed at a group of subscribers interested in a particular broadcast.

These two opposed modes present different security and technical characteristics as well as different loads to the IPTV infrastructure.

Unicast is used to allow subscribers to have access to a specific data stream that has been specifically requested by the set top box and is being sent by the video-on-demand server. Unicast traffic is not recommended for a large number of users as it overloads the network. IPTV service providers must allocate large amounts of network and capacity to support unicast streams.

For a large number of subscribers interested in the same stream, IPTV service providers can use multicast sessions to broadcast standard TV programs.

Multicast traffic has some limitations specifically related to the UDP protocol used for delivery. It is incapable of detecting packet loss and has no native mechanisms to react to network congestion. Network elements use queues to manage multicast traffic, using priority identifiers to manage the traffic.

Figure 3.7 illustrates the unicast traffic from the video-on-demand server (or the video streaming server) towards set top boxes. Each one of the arrows represents a different stream.

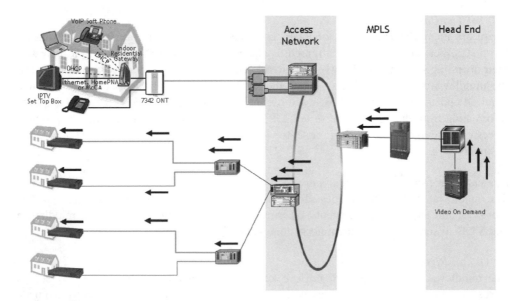

**Figure 3.7**  Unicast Traffic on IPTV

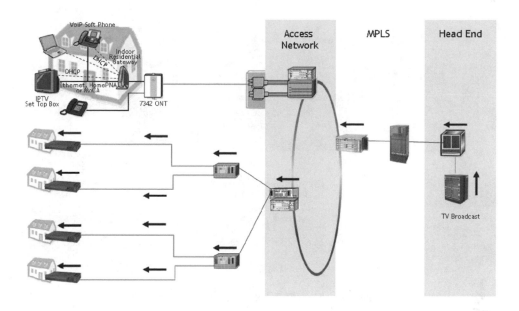

**Figure 3.8**  Multicast traffic on IPTV

Individual streams are sent to each set top box and will require dedicated network bandwith and processing resources at the head end.

Figure 3.8 illustrates the multicast traffic broadcast by the head end (video streaming server). A single stream is sent by the broadcast server and then individual switches and routers will redistribute the same stream to those set top boxes that have subscribed to the broadcast domain. The bandwith requirements and processing power demands are much lower than in unicast environments.

The transport and aggregation network has different mechanisms to ensure secure and reliable delivery of streams to subscribers. One of the mechanisms used for this process is the use of virtual local area networks (VLANs). These are used to segment and isolate the traffic, avoiding unauthorized access to contents and facilitating the flow control.

Between the head end and home end there are a number of VLANs dedicated to specific traffic. Figure 3.9 illustrates some of the VLANs used.

It is important to note that, in almost all scenarios, the transport and aggregation network will be shared between different applications including high-speed Internet access (HIS), voice-over IP (VoIP), IPTV applications and other functions offered by the network provider. The network elements will be able to add VLAN tags to the traffic coming from different locations, transport it through the network and remove the tag when necessary.

VLANs are based on the IEEE 802.1Q/P standards [3]. The following figure shows a typical Ethernet frame with the optional VLAN tag, which includes a VLAN identifier (a 12-bit field identifying the community to which a particular packet belongs). VLANs become Ethernet broadcast domains, with network elements releasing packets only to members of a particular domain.

The basic fields of a packet used for VLANs under the IEEE 802.1q are presented in Figure 3.10.

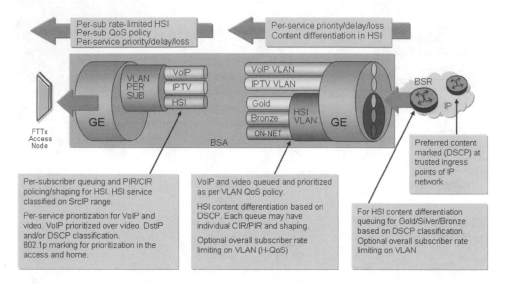

Figure 3.9    View of VLANs on aggregation network

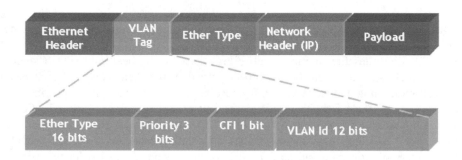

**Figure 3.10**    VLAN packet as per IEEE 802.1Q

Within IPTV environments, the VLANs typically found are customer VLANs (C-VLAN) or multicast VLANs (MC-VLAN) depending on the type of traffic required. MC-VLANS are typically used for broadcast TV streams, set top box updates and other applications that are intended for a large audience.

Set top boxes will send their local DSLAM a request to join a particular MC-VLAN. If the DSLAM is not already receiving the stream, it will request it from the head end and, once it is received, will confirm to the set top box that it has joined the MC-VLAN. Multicast traffic is supported by the Internet group membership protocol (IGMP) [4]. Set top boxes use IGMP either to join or to leave particular streams. These commands are sent to the DSLAM or aggregation router where the validation and aggregation from various subscribers are undertaken.

There are two main participants in the IGMP traffic: IGMP routers and IGMP hosts.

IGMP routers receive the join and leave requests from hosts and determine if a particular stream should be sent. This can be validated further using business rules. The router will update traffic-related information using a routing protocol or static information.

IGMP hosts send requests to the IGMP routers in order to leave or join a particular stream. The requests can be to JOIN, LEAVE or QUERY the particular state of the communication. In cases where there is inconsistency in the communication, the set top box can issue a QUERY packet to establish the list of streams being sent to it. Each TV channel is a particular IP multicast group being sent within the MC-VLAN.

IGMP allows DLSAMs and basically the IPTV network infrastructure to determine which set top boxes should receive streams. It is important to note that, in spite of the fact that traffic is being broadcast, this does not imply that all set top boxes will receive all streams. In a satellite or cable TV service, all end-elements will receive the complete set of broadcast channels at all times. This increases the risk, as set top boxes will have constant access to contents (even if they are encrypted), which simplifies the process of breaking the encryption and protection mechanisms. In a properly configured IPTV environment, set top boxes will not receive streams to which they are not entitled. If, for some reason, subscribers have been able to lift the security controls at the set top box or have obtained stolen DRM keys enabling them to unscramble all contents, the DSLAM will still block those streams to which the subscribers are not entitled.

The basic structure of an IGMP v2 packet, including some references to the expected sizes of fields, is shown in Figure 3.11.

The structure of the IGMPv2 packet includes the type, maximum response time, IGMP checksum and group address.

The type can take one of the options listed in Table 3.10, which is based on the IANA assignment [5].

The maximum response time field is used within the membership query messages and it defines the time before a response must be sent. Changes to this value allow for increasing or reducing the levels of traffic when hosts leave a multicast group.

The checksum is used for validation and protection against errors.

The group address is used to identify each multicast group. When sending a type $0 \times 12$ membership report or a type $0 \times 17$ leave group message, this field includes the IP multicast

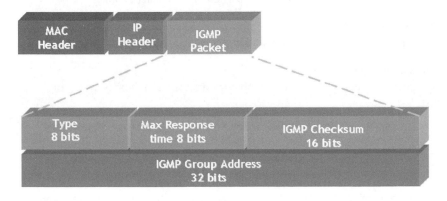

**Figure 3.11**  IGMPv2 packet structure

**Table 3.10**  IGMP v2 type options

| Type | Name | Reference |
|------|------|-----------|
| 0 × 11 | IGMP membership query | [RFC1112] |
| 0 × 12 | IGMPv1 membership report | [RFC1112] |
| 0 × 13 | DVMRP | [RFCDVMRP] |
| 0 × 14 | PIM version 1 | [PIMv1] |
| 0 × 15 | Cisco trace messages | |
| 0 × 16 | IGMPv2 membership report | [RFC2236] |
| 0 × 17 | IGMPv2 leave group | [RFC2236] |
| 0 × 1e | Multicast traceroute response | |
| 0 × 1f | Multicast traceroute | |
| 0 × 22 | IGMPv3 membership report | [RFC3376] |
| 0 × 30 | Multicast router advertisement | [RFC-ietf-magma-mrdisc-07.txt] |
| 0 × 31 | Multicast router solicitation | [RFC-ietf-magma-mrdisc-07.txt] |
| 0 × 32 | Multicast router termination | [RFC-ietf-magma-mrdisc-07.txt] |
| 0 × f0–0 × ff | Reserved for experimentation | [RFC3228, BCP57] |

group address in question. On a type 0 × 11 membership query this field is 0 when sending a general query, and it has the value of the group address when sending a group-specific query.

The commonly used packets within IGMPv2 are as follows:

*(i) Membership Report*
Type 0 × 16 membership reports are used by the set top box to JOIN a broadcast channel or to respond to a solicited membership query from the DSLAMs and/or multicast routers.

*(ii) Leave Group*
A type 0 × 17 leave group is used by the set top box to leave a particular multicast group.

*(iii) Membership Query*
A type 0 × 11 membership query is sent by the DSLAM or multicast router to establish whether any set top boxes are receiving a particular broadcast channel.

*(iv) General Query*
A general query is used to ask if any set top box is receiving any broadcast.

*(v) Group-specific Query*
A group-specific query is used to establish if any set top box is receiving a particular broadcast.

Set top boxes must follow the IGMPv2 registration process in order to join a particular multicast channel. Figure 3.12 shows an example of the traffic flow between the set top box and the DSLAM or router.

When the subscriber wants to join a particular broadcast stream, the set top box will send an unsolicited membership report addressed to the group it wants to receive. Once the multicast router receives the request, it will start sending the stream to the specific interface for the set top box as long as it is authorized.

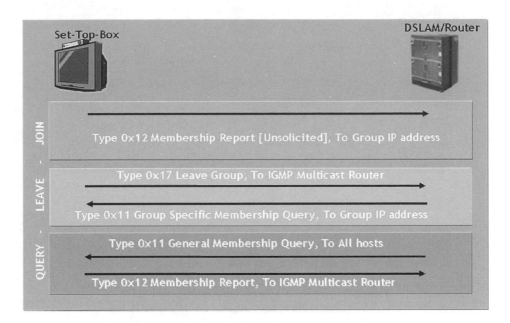

**Figure 3.12**   IGMPv2 traffic example

If the subscriber wants to leave the channel, the set top box will send a leave group message to the multicast router address including the group address of the particular stream. The router will then send a group-specific query to establish if there are any set top boxes left receiving the stream. This request will have a maximum response time value for their response to be sent, and, if no response is received within the window, the router will stop forwarding the stream on that particular interface.

A general membership query is sent periodically to all set top boxes to establish which streams are being used. Maximum response time values are set on this query as well. This is a mechanism that can be used by the IPTV infrastructure to establish the current state of the network and reduce the number of streams being forwarded.

IPTV increases the bandwidth demand. A standard-definition channel will require close to 3 Mbps, and high definition could demand 6–10 Mbps. The number of channels available is also increasing rapidly, and unicast demands are also increasing. An alternative to some of these pressures was the use of some of the new options defined within the IGMPv3, including the source-specific multicast (SSM), also referred to as the single-source multicast. As a security mechanism, the use of SSM reduces the chances of peer-to-peer traffic, specifically, subscribers communicating directly with other subscribers on the same multicast stream. Spam, phishing and similar threats are reduced, as well as the effect of computer worms and viruses infecting set top boxes or computers at the home end.

### *IGMPv3*
There are two main multicast messages described by RFC 3376 [6]: the type $0 \times 11$ membership query and the type $0 \times 22$ version 3 membership report.

Figure 3.13 shows a simplified view of an IGMPv3 membership query packet (RFC 3376) with the expected field sizes.

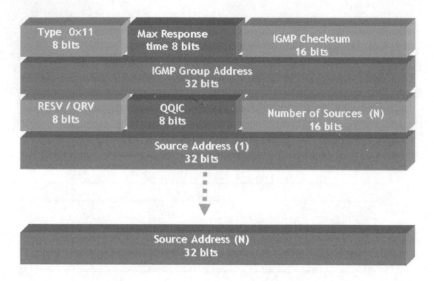

**Figure 3.13**  IGMPv3 membership query – RFC 3376

*(i) Membership Query Message*

Type $0 \times 11$ membership query messages are sent by multicast routers to enquiry about the state of interfaces and follow a standard format:

- Type field: for membership query messages the value is $0 \times 11$;
- Max. response code: the maximum allowed time after a query has been sent to a broadcast domain;
- Checksum: used to validate the integrity of the message;
- Group address: the value of this field is zero within a general query and the IP multicast address is used when sending a group-specific query or group- and source-specific query;
- Reserved: reserved for future use;
- QRV (Querier's Robustness Variable): used by multicast routers to determine the robustness of queries;
- QQIC (Querier's Query Interval Code): querier's interval used by the querier;
- Number of sources $N$: this value defines how many source addresses are included in the query; it is zero in a general query or a group-specific query and nonzero in group- and source-specific queries;
- Source address: the source address of the multicast stream.

There are three main query options within IGMPv3:

1. *General query.* Used by a multicast router to establish the reception state of the network. For this query the group address and number of sources fields are zero.
2. *Group-specific query.* Used by a multicast router to establish the state of a single multicast address. For this query the group address field includes the specific multicast address queried. The number of sources value is zero.

3. *Group- and source-specific query.* Used by multicast routers to establish if neighboring interfaces want packets sent to a specified multicast address from any of the available source addresses. The group address includes the multicast address queried, and the source address field includes the source address of interest.

*(ii) Type 0 × 22 Version 3 Membership Report Message*
The version 3 membership reports are sent by IP systems to communicate their present state or changes to the state. The reports follow a standard format.

Figure 3.14 shows a simplified view of an IGMPv3 membership report packet (RFC 3376) with the expected field sizes.

Figure 3.15 shows a simplified view of an IGMPv3 membership report (group structure) with the expected field sizes.

- Reserved: this field is available for future use;
- Checksum: used to control the integrity of the message;
- Number of group records $M$: this field defines the number of records on the report;
- Group record: includes a number of fields stating the sender's membership to multicast groups;
- Aux. data length: the length of the auxiliary data field;
- Number of sources $N$: states how many source addresses are included in the group record;
- Multicast address: the multicast address to which the record refers;
- Source address: this includes a number of unicast addresses related to the number of sources;
- Auxiliary data: not used by IGMPv3;
- Packet flow on IGMPv3: when available, set top boxes can use IGMPv3 and its registration process in order to join a particular multicast channel; Figure 3.16 shows an example of the traffic flow between the set top box and the DSLAM or router;

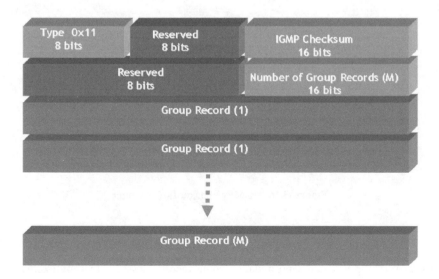

**Figure 3.14**  IGMPv3 membership report – RFC 3376

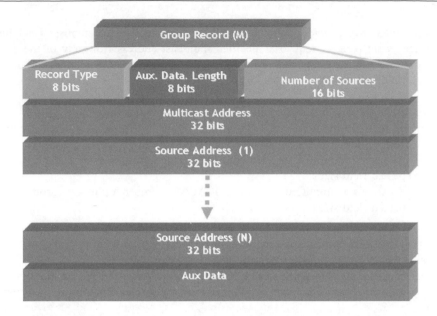

**Figure 3.15**   IGMPv3 membership report – group structure

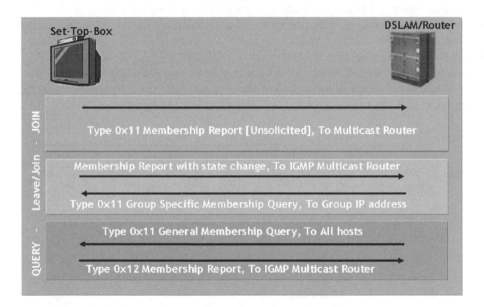

**Figure 3.16**   IGMPv3 packet flow example

- JOIN: equivalent to IGMPv3, the multicast router will add the set top box address to the multicast group list;
- LEAVE: a set top box can send a state change record message, thereby excluding itself from the streams it no longer wishes to receive; the router will query the multicast

domain to establish if other set top boxes are interested in the stream or if it is no longer required;

• QUERY: there are frequent general membership queries to determine which set top boxes are available.

*IGMP Proxy*

DSLAMs are often used either to bridge IGMP traffic (snooping) or as an IGMP proxy receiving all IGMP requests and sending them upstream. When acting as an IGMP proxy, all set top boxes will see the DSLAM as the IGMP multicast router. Once received, the IGMP requests will be sent by the DSLAM upstream acting as a client.

### 3.3.2.1 RP and RTSP

Protocols supporting unicast traffic from head end to set top boxes and VCR-like functions to stop, forward and pause streams.

*RTP*

Real-time transport protocol is used for end-to-end transport of video streams within IPTV. It can be used over both unicast and multicast environments. Defined by RFC 3550 [7].

RTP is not involved in resource allocation or quality of service – those functions will be provided by other elements within the infrastructure, including the DSLAM and switches.

Data transport uses RTCP to allow monitoring of data delivery and some identification functions.

There are no standard ports assigned to RTP, but there are some commonly used ranges: 16 384–32 767, gnome meeting; 5000–5003 and 5010–5013, real-time transport; 6970–6999, RTP ChatAv; 16384–16403 and 16384–32767, RP.

RTP is used for payload-type identification, sequence numbering and time stamping.

The standard structure of the RTP packets as defined by RFC 3550 is depicted in Figure 3.17.

### 3.3.2.2 RTSP [8]

*(i) Port Numbers*
The following ports have been registered with IANA:

• rtsp 554/tcp real-time stream control protocol;
• rtsp 554/udp real-time stream control protocol;
• rtsp-alt 8554/tcp RTSP alternate (see port 554);
• rtsp-alt 8554/udp RTSP alternate (see port 554).

**Figure 3.17**   RFC 3550 RTP – real-time transport protocol

*(ii) RTSP Commands*

The protocol is similar in syntax and operation to HTTP, but RTSP adds new requests. While HTTP is stateless, RTSP is a stateful protocol. A session ID is used to keep track of sessions when needed. In this way, no permanent TCP connection is needed. RTSP messages are sent from client to server, although some exceptions exist where the server will send to the client. A number of typical HTTP requests, like an OPTION request, are also frequently used.

Some of the commands used by RTSP are as follows:

- *Describe.* This method requests information about a particular object. This is used to obtain the necessary parameters to have access to the video contents.
- *Announce.* This is used by the set top box to provide a description of the object to the server, or by the VOD server, to provide updates on the session.
- *Set-up.* This is used by set top boxes to request changes on the transport parameters. If a networking element (such as firewalls or stateful inspection routers) is involved in the communication, it will be able to use the set-up parameters to understand the request and socket to be used. The set-up also starts the RTSP session.
- *Play.* This method is used by set top boxes to instruct the VOD server to start sending the stream using the parameters included in the set-up request following confirmation. The position used for the stream is the start of the range available and will deliver a stream until the end of the content. The play and other methods have a similar structure to HTTP. Web application firewalls, reverse proxies and IDS/IPS systems may be able to detect attacks sent to the VOD server by set top boxes using modifications to the expected RTSP values:

PLAY rtsp:<filename> RTSP1/.0
CSeq: 3
Range: npt = 0–
Scale: 1
Session: 123456789101112131415

- *Pause.* This stops the stream leaving a socket dedicated to the particular set top box.
- *Teardown.* In contrast to pause, the teardown method will release the resources allocated for a particular set top box; the session is removed from the system.
- *Get parameter.* This method is used by set top boxes to obtain parameters from a specific object.
- *Redirect.* This is used by the VOD server to instruct set top boxes to connect to a different server. This request informs the client that it must connect to another server.
- *Record.* This is used by set top boxes to request the recording of a particular stream and the storage of the subsequent file in a particular location. The record method may include start and end parameters or by default would be the whole object.

Set top boxes obtain the RTSP session description from the VOD web server. They will then send RTSP commands to the media server, allowing for setting up the session, playing, pausing and finalizing the session. This traffic is depicted in Figure 3.18.

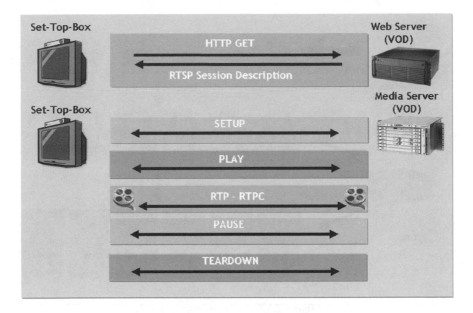

**Figure 3.18**   RTSP operation – example

### 3.3.2.3 Ismacryp

The Internet streaming media alliance (ISMA) developed the ISMA encryption and authentication specification (Ismacryp) [9]. The standard covers encryption and authentication of content that is streamed over Internet protocol and is intended to be independent of media players, DRM systems, key management schemes, etc. The standard supports a number of codecs in addition to MPEG-4. The underlying encryption algorithm used is the advanced encryption standard (AES).

Traditional conditional access systems are efficient at protecting traffic on unidirectional links on IP networks. However, the nature of IPTV and the type of flexibility and controls provided to subscribers create a number of complex issues. For example, scrambled content would be very difficult to manage using VCR-like controls as encryption keys may be linked to future content and as such may not be available.

Ismacryp supports bidirectional links by encrypting the stream and then creating packets with the content. The packets can be stored and made available through other mechanisms.

Figure 3.19 illustrates the Ismacryp concept of taking the original content and dividing it into small pieces to be encrypted and stored for future use. Access to the encrypted sections will be done using RTP.

With this approach, content is encrypted at the source and it can either be stored or streamed depending on the specific needs. This adds flexibility to the transport and distribution of the content as well as the type of medium used for distribution to end-users.

IPTV service providers can both decrypt the content at the head end and protect the content with their own DRM application, or they can add a second layer of encryption to the content. This approach is compatible with different key management systems (KMSs), adding flexibility to the solution.

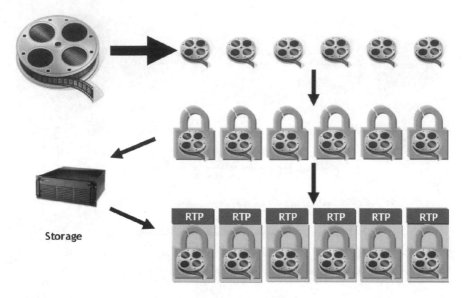

**Figure 3.19**  Ismacryp approach

### 3.3.2.4  PIM

PIM stands for protocol-independent multicast [10]. It is used for requests within the IPTV network as well as requests for multicast data forwarding. Commonly used multicast routing protocol suite comprising PIM-DM, PIM-SM and PIM-SSM.

*(i) PIM-DM*
Protocol-independent multicast – dense mode [11]. Used on multicast LAN applications. It uses the routing table populated by any underlying unicast routing protocol to perform reverse path forwarding (RPF) verifications. PIM-DM can use the unicast routing table populated by OSPF, IS-IS, BGP, etc.

*(ii) PIM-SM*
Protocol-independent multicast – sparse mode. PIM-SM uses a rendezvous point (RP) which multicast providers use to register sessions. The RP maintains a table with source and group information. When a particular host requires access to a specific multicast session, it sends a join request to the gateway for a multicast group. The router will build a path to the RF who has the entry table.

This approach is known as the IP multicast model of receiver-initiated membership, supports both shared and shortest-path trees and uses soft-state mechanisms to adapt to changing network conditions.The RP will build a tree from the destination back to the source, and it will forward multicast packets to the destination. When packets are received from the source, the gateway can build a path directly without relying on the RP.

*(iii) PIM-SSM*
Protocol-independent multicast – source-specific multicast [12]. PIM-SSM does not rely on the use of RPs. It is based on a one-to-many approach, compatible with IPTV and specifically

with TV channel distribution. It is used to create shortest-path trees (SPTs), and the router closest to the destination receives information of the unicast IP address of the source for the multicast traffic.

### 3.3.2.5 MSDP

Multicast source discovery protocol [13] is used to connect shared trees without using interdomain shared trees. MSDP supports PIM-SM, PIM-DM and other protocols. MSDP uses independent RPs, avoiding dependencies on external RPs and increasing protocol flexibility.

This protocol allows domains to discover sources from other domains. MSDP shares the information on sources for a particular group to be advertised to all RPs, using source-active (SA) packets. SA packets include source and group information for RPs. This information must be filtered to avoid false packets being distributed.

### 3.3.2.6 DSM-CC

The digital storage media command and control (DSM-CC) [14] is used to implement control channels for video streams, particularly for VCR-like controls (fastforward, rewind, pause, etc.). DSM-CC works with protocols found on IPTV networks such as RSVP, RTSP and RTP. The standard is based on a three-component model: client, server and session and resource manager (SRM). The SRM allocates and manages network resources (such as channels, bandwidth and network addresses.)

The main scenarios for this protocol are:

- user–network client configuration;
- user–network session protocol;
- user–user directory, stream control, file access;
- interactive and broadcast download;
- broadcast object carousel;
- switched digital broadcast channel change protocol.

### 3.3.2.7 Internet Service Provider

Set top boxes require both IP addresses and domain name resolution to be able to operate. This data is provided by the DHCP servers and the DNS servers, both commonly used for the high-speed Internet access functions and standard services offered by the ISP.

These services are critical, and they must be properly secured to avoid denial of service attacks and spoofing of subscriber accounts.

### 3.3.2.8 DSLAM

The digital serial line access multiplexer (DSLAM) has a number of DSL modems receiving the sessions from subscribers. It then brings all the sessions together out on a backbone connection to the aggregation network.

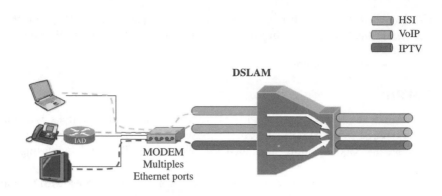

* The Permanent Virtual Circuit *is linked with specific VLANs*

    * One PVC for High Speed Internet access

    * Additional PVC's for IPTV, VoIP and management

**Figure 3.20**   DSLAM capabilities

The DSLAM is one of the last borders completely managed by the IPTV network providers as it is hosted within the secure physical environment, and, in contrast to the set top box and residential gateway, subscribers have no physical access to the DSLAM.

Figure 3.20 illustrates how the DSLAM facilitates the high-speed Internet, VoIP and IPTV VLANs to subscribers. Equipment at the home end will split the traffic and allow VLANS to be terminated (it can also be terminated at the DSLAM).

Modern DSLAMs are able to aggregate all modem sessions into one signal and direct this via TCP/IP (or other protocols) to the transport network. In some models there are networking options such as IGMP proxy functions, VLAN and virtual circuit functions as well as a number of TCP/IP applications that support the operation of IPTV networks. DSLAMs use the characteristics of voice related to using relatively low frequencies of the spectrum and allocate the high frequencies for data communications. In this way, both voice and data can be managed by DSLAMs.

DSLAMs use Ethernet bridging (VLANs). This works by marking the incoming packets with a VLAN-ID which represents a group. VLAN-IDs are based on the 802.1Q and are used by DSLAMs to undertake the forwarding process. On the basis of the internal forwarding table, the DSLAMs will send packets to different ports, always separating traffic from different VLANs.

Each virtual bridge then performs an independent verification of the traffic.

The upstream traffic coming from a particular port/PVC is assigned to a particular VLAN. These packets are forwarded to a particular Ethernet port following the forwarding table on the DSLAM. Upstream traffic from different set top boxes is aggregated on VLANs dedicated for particular services.

Figure 3.21 illustrates the general view of the DSLAM and how VLANs are terminated at the BRAS, head end or at the Internet components. Several DSLAMs will share access to a single VLAN.

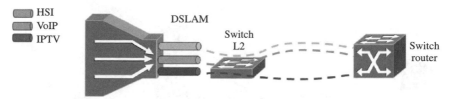

- Ethernet Network
    - SVLAN with PPPoE traffic terminates at the BRAS(NISIP)
    - The rest of the VLANs are directed to specific services, like the video head end
    - Video VLAN and voice VLAN are shared by many DSLAMs
- Functionalities
    - DSLAM with DHCP relay and option 82
    - DSLAM with Multicast and IGMP (snooping or proxy)
    - DSLAM acting like bridge
    - SVLAN ('selective queue') at the DSLAM

**Figure 3.21**   General vision of DSLAM VLANs

Set top boxes require an IP address to join the network. Their configuration is managed using the DHCP protocol. DHCP will also be supported by authentication mechanisms (AAA systems) to enable control by the middleware and business applications on the type of access provided to the subscriber.

DSLAMs have a DHCP relay agent that inserts the physical line identifier (option 82 on DHCP) on the address request. This ensures that messages coming from the set top box are linked with a particular physical location, reducing the chances of spoofing and fraud. The authentication and accounting process relies on option 82 to validate users, and, using the RADIUS service, it is possible to ask business applications what type of access should be granted to the subscriber (or what IP address/VLAN). DSLAMs can also use IEEE 802.1X authentication, using RADIUS to support the network authentication process. This is done using extensible authentication protocol (EAP) between the set top box and the authentication servers. The DHCP server will wait for the middleware server to authorize any IP address to be assigned to a set top box.

Figure 3.22 illustrates the set top box authentication process. It starts with a DHCP request issued by the set top box and forwarded by the DSLAM and routers to the DHCP server. The request includes physical line information on option 82 of the DHCP request. The DHCP server verifies with other components such as the RADIUS or middleware server if the physical line and subscriber should receive a valid IP address, and, once confirmation is obtained, the access is granted.

DSLAMs participate in the multicast process by providing either snooping or proxy functions for IGMP requests. Upstream IGMP traffic will be VLAN tagged to ensure segregation of traffic on the aggregation network. IGMP messages are snooped by bridges to optimize the multicast distribution tree. The video stream will be duplicated to all relevant ports as per the IGMP join requests received by the DSLAM from set top boxes. Video streams are encrypted, so set top boxes will require the DRM keys as well as authorization to join a particular IGMP domain.

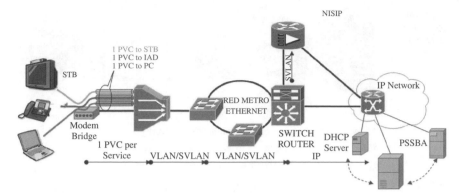

- The server must verify with the middleware that this client has access to the video services, based on option 82 of the DHCP request packet.
- The DHCP server has to verify with the PSSBA that this client is enabled to use the sercice.

**Figure 3.22**   Set top box authentication

To provide antispoofing and enhanced security, the DSLAM will check the source ip address of each subscriber for each request against the list of provisioned subscribers. A mapping of the ip address and its corresponding physical port is stored into the DSLAM. The source ip address of every upstream packet entering the DSLAM via the subscriber port is verified against the table. DSLAMs also provide reports on user behavior. All requests from set top boxes are logged. Figure 3.23 illustrates the multicast architecture.

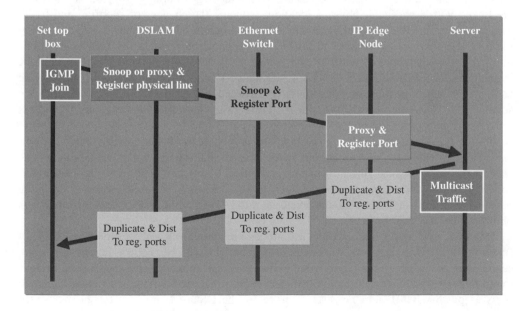

**Figure 3.23**   Multicast architecture – DSLAM participation

**Table 3.11**  DSLAM asset value

| Asset – DSLAM | |
| --- | --- |
| Relative Value | Exposure Level |
| The DSLAM tends to support between 2000 and 6000 set top boxes. This number is relatively small compared with the overall population of subscribers. An individual DSLAM is not considered to be a high-value asset. Intruders do not gain immediate benefits from manipulating a DSLAM. Content streams stolen from the DSLAM cannot be used without the DRM keys. | The exposure level of DSLAMs is relatively high because they have broadband connection to the set top boxes. However, most types of DSLAM are transparent to the network and do not offer services to the subscribers. |

DSLAMs also provide a flexible definition of channel bundles. The creation of channel bundles (a number of TV channels in a set) allows a flexible assignment of subscribers. Modification of the bundles (for example by adding a TV channel to a bundle) can be done without the need for rearrangement of user-channel bundle assignment. Table 3.11 presents the relative value and exposure level of the DSLAM.

### 3.3.3 Home End (Subscriber)

The home end includes a network termination device, basically the access point from the network. The termination device will be connected to a modem that will convert the information to IP based, and in some cases a splitter will be used to provide voice services if public service telephone networks are used.

A gateway will be used to separate the IP services (data, video, voice), and in some cases the gateway includes a firewall, DHCP service and other networking services required to improve the service.

Customers require a set top box that in most cases is provided by the service provider. In some cases a home PC will be used to connect directly to the network and an STB is not required. Additional elements include a phone terminal for either PSTN or VoIP and wireless access routers.

The home end is out of scope from the security mechanisms established by the IPTV service provider, and it should be assumed that any element hosted within the home end will eventually be modified or bypassed by subscribers. There are many examples of cable and DSL modems being modified and sold on the black market. These modifications allow untapped bandwidth, and similar situations could happen to the set top box.

Although some elements sit between the DSLAM and the TV set, it can be assumed that traffic will be safe until it reaches the set top box. It is advisable to liaise with the IPTV network provider to ensure that the residential gateway (RG) has been properly configured to avoid unauthorized access to the home end and not allow unauthorized connections to the RG operating system.

### 3.3.3.1 Set Top Box

The IP-based set top box is used to connect the IPTV head end with the TV set. The main function of this element is to interpret and translate the requests from the subscribers and send IP-based messages to the head end, requesting specific contents or services. The set top box will receive encrypted contents and will have to decrypt and decode them to be presented to the set top box.

Figure 3.24 illustrates the typical IP set top box architecture. It includes references to the most typical components on modern set top boxes. Many alternative models can be found with similar basic functions. New developments of the technology would clearly bring new elements within set top boxes.

The typical components found on an IP set top box are as follows:

*(i) Hardware CPU*
IP set top boxes tend to have low-end chipsets. The processing power and memory is limited compared with standard PCs. Manufacturers tend to select basic CPUs that will provide enough processing power to deal with the basic functions and appropriate response times.

*(ii) Core System – Secure Crypto*
The core hardware includes the different electronic components supporting the operation of the IP set top box, information exchange between components, memory and most importantly the possibility of a dedicated chip to store the DRM and PKI keys required for access and authentication. With the use of a dedicated chip for storing keys, the risks of unauthorized access are reduced, as opposed to a software-based solution that stores all security elements on a hard drive or ROM component.

*(iii) Peripherals*
There are a number of peripherals connecting to the set top box, including the network cable, video output (in different formats), infrared component for the remote control or keyboard, USB and storage technologies.

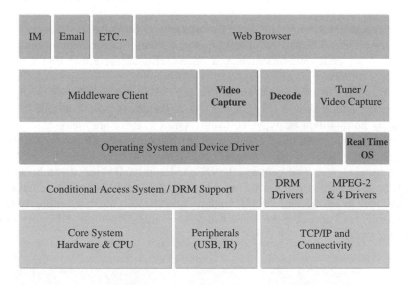

**Figure 3.24**   Typical IP-STB architecture

### (iv) DRM and Conditional Access System

The set top box requires a dedicated component to deal with DRM-related functions. It is necessary to request and update the DRM keys, decrypt the content and provide the encoded stream to other components. Additionally, the set top box needs to authenticate itself against the conditional access system to enable content access. The IPTV service provider must ensure that the appropriate DRM and CAS applications are loaded onto the set top box. This specific requirement makes it very difficult for an open market on IPTV services. It is unlikely that subscribers would be able to switch IPTV service providers without changing the set top box. Additionally, it is unlikely that a content aggregator could operate in the market offering contents from a large number of content owners or IPTV service providers, as each service provider may have a different DRM system. Over time, standards will be developed to ensure interoperability between DRM and CAS systems. Meanwhile, subscribers will be linked to their own IPTV service provider.

### (v) MPEG 2 and MPEG 4 Drivers

The set top box needs a number of MPEG-2, MPEG-4 and, in general, codec drivers to be able to decode the stream and provide an output that can be displayed by the TV set. In general, IPTV service providers must ensure that appropriate codecs are loaded into the set top box. The codecs used on the head end should be loaded on the set top box.

### (vi) Operating System and OS Drivers

Lightweight operating systems are used for set top boxes. Some open-source and proprietary operating systems are used for this function. One of the advantages of these operating systems is the flexibility they offer for developers of the set top box platform, as they include the browser support as well as the email, instant messaging and networking support. Being a standard operating system, there are clear risks of worms and viruses affecting the operation of the set top box. It is important that the master copy of the operating system is properly configured, patches applied and known vulnerabilities fixed before releasing the set top box to subscribers. Additionally, all unnecessary ports must be blocked.

### (vii) Middleware Client

A specific client that communicates with the middleware server. This client may use the web browser to exchange information with the middleware server as well as download the EPG to be displayed to the subscriber. The middleware client may include DRM functions in some cases.

### (viii) Video Capture – Decode

This function will receive the output stream from the DRM function and will decode the MPEG-4 data into a usable format. This may include decoding the data into NTSC/PAL output.

### (ix) Web Browser

The middleware servers tend to operate as web services. Some implementations will provide all access using SSL (HTTPS, port 443). The web browser is used by the set top box to access the contents and display information to the subscriber. Some set top boxes may allow interaction by the subscriber with the set top box and enable web browsing of Internet sites by the subscriber.

*(x) Instant Message Client*

Some set top box implementations may include an instant messaging client and other similar functions such as caller ID and instant quotes.

*(xi) Email Client*

Set top box vendors may include a basic email client to allow subscribers to send and receive messages. This application may include viewers for typical files such as documents, spreadsheets and presentations.

Figure 3.25 illustrates the IP set top box process flow, specifically starting with IP requests and reception of content, web browser functions, middleware interaction and content decryption, following with stream decoding and encoding for NTSC/PAL or the relevant standard and ultimately providing the output for the TV set. Table 3.12 presents the relative value and exposure level of the set top box.

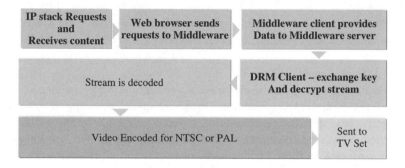

**Figure 3.25**   Typical IP-STB Process

**Table 3.12**   Set top box asset information

| Asset – Set Top Box | |
| --- | --- |
| Relative Value | Exposure Level |
| The set top box holds a number of encryption keys that can be used to decrypt content. Additionally, it can be used to request premium content from the head end. Intruders who control one or several set top boxes could be able to steal and copy content without having to break into the head end. In general, the set top box is an extension of the content repository. Understanding that the set top box is outside the control of the IPTV service provider, it is important to use hardware security mechanisms to avoid theft of content. | The exposure level of the set top box is high. Not only is it outside the physical control of the IPTV service provider, it is also based on standard operating systems with known security problems. |

## 3.4 Summary

The IPTV environment can be represented by four main components on the high-level architecture: content provider, IPTV service provider, network provider and subscriber.

A functional architecture can be defined with more details about the components including: IPTV control, content delivery, content provisioning, IPTV transport and subscriber. These functional components can be assigned to different teams, ensuring segregation of duties and proper specialization.

Functional components agglutinate components with similar functions, although in some situations a component may be placed in two categories. Components can be assigned to different functional components. For example, the set top box would be added to the subscriber functions and content provisioning would include the satellite receivers.

Some of the critical components of the head end are as follows.

*(i) Satellite Receivers*
Integrated receiver decoder (IRD).

*(ii) Video Repository*
This includes:

- video library;
- media library;
- library servers;
- storage area network;
- video-on-demand movie database;
- film server (video and audio files).

*(iii) Content Management System*
This includes:

- command center;
- asset management system;
- digital rights management.

*(iv) Master Video Streaming/Game Server*
This includes:

- propagation service;
- streaming service.

*(v) Ingest Gateway (Video Capture)*
This includes:

- recording system;
- recording manager;
- capture/distribution server.

*(vi) Video Cache Streaming Server*
This includes:

* caching server;
* media cluster.

*(vii) Middleware*
Middleware servers.

*(viii) Business-related Systems*
* accounting;
* provisioning;
* customer information.

Some vendors will be able to merge several functions on the same server. For example, the asset management server and video repository may be merged into a single server, but functions and access requirements may be different.

Networking equipment provides critical functions such as aggregating the requests from thousands of set top boxes and liaising with network components to authenticate set top boxes and assign valid IP addresses.

# References

[1] ATIS-0800007, '*IPTV High Level Architecture*', ATIS-IIF, 2007.
[2] Chiariglione, L., '*MPEG-2 Generic Coding of Moving Pictures and Associated Audio Information*', 2001. Available online: http://www.chiariglione.org/MPEG/standards/mpeg-2/mpeg-2.htm [2 October 2007].
[3] Institute of Electrical and Electronics Engineers, '*802.1Q – Virtual LANs*'. Available online: http://www.ieee802.org/1/pages/802.1Q.html [2 October 2007].
[4] Fenner, W., '*Internet Group Management Protocol, Version 2*', IETF, 1997. Available online: http://tools.ietf.org/html/rfc2236 [2 October 2007].
[5] IANA, '*IGMP Type Numbers - per [RFC3228, BCP57]*', 2005. Available online: http://www.iana.org/assignments/igmp-type-numbers [2 October 2007].
[6] Internet Engineering Task Force, '*Internet Group Management Protocol, Version 3*', 2002. Available online: http://www.ietf.org/rfc/rfc3376.txt [2 October 2007].
[7] Internet Engineering Task Force, '*RTP: a Transport Protocol for Real-Time Applications*', 2003. Available online: http://www.apps.ietf.org/rfc/rfc3550.html [2 October 2007].
[8] Internet Engineering Task Force, '*Real Time Streaming Protocol (RTSP)*', 1998. Available online: http://www.ietf.org/rfc/rfc2326.txt [2 October 2007].
[9] Internet Streaming Media Alliance, '*Encryption and Authentication, Version 1.1*', 2005. Available online: http://www.isma.tv/technology/ISMACryp1.1.html [2 October 2007].
[10] Internet Engineering Task Force, '*Protocol Independent Multicast (PIM)*', 1998. Available online: http://www.ietf.org/rfc/rfc2326.txt [2 October 2007].
[11] Internet Engineering Task Force, '*Protocol Independent Multicast (PIM) Dense Mode*', 1999. Available online: http://tools.ietf.org/id/draft-ietf-pim-v2-dm-03.txt [2 October 2007].
[12] Internet Engineering Task Force, '*Protocol Independent Multicast (PIM) Source Specific Multicast*', 2003. Available online: http://www.ietf.org/rfc/rfc3569.txt [2 October 2007].
[13] Internet Engineering Task Force, '*Multicast Source Discovery Protocol (MSDP)*', 2003. Available online: http://www.ietf.org/rfc/rfc3618.txt [2 October 2007].
[14] International Organization for Standardization, '*DSM-CC*', 1997. Available online: http://www.chiariglione.org/mpeg/faq/mp2-dsm/mp2-dsm.htm [2 October 2007].

# 4

# Intellectual Property

From a business perspective, IPTV service providers rely on both subscribers and advertisers to provide viable revenue streams. Also involved are the content owners who assign distribution rights to IPTV service providers in very specific regions.

IPTV service providers must ensure that contents are enjoyed by subscribers under the terms of the licenses assigned by content owners. This is accomplished using digital rights management technologies. DRM is used to apply the intellectual property restrictions on digital assets.

This chapter will present the history and evolution of intellectual property (IP) as well as the existing technologies that can be used within IPTV environments to ensure the protection of digital assets. DRM is heavily supported by encryption technologies. Readers may want to expand their understanding of encryption algorithms and techniques before undertaking a review of live DRM systems.

## 4.1 Introduction

Intellectual property is the term used to describe the legal rights protecting creative works, inventions and commercial goodwill. Intellectual property rights are used to discourage the theft or use without permission of a particular creative work while providing the legal support for different business models to be developed around intellectual property assets. This is based on the principle that the individual who has made an intellectual effort to create an asset must be able to benefit from this effort. This principle also applies, in an equivalent manner, to an individual who has made a manual effort physically to create an asset.

IP rights protect the large investments made by music studios, movie studios and the entertainment industry in general and help to create a cycle of revenues that would support new and improved materials while providing profits to shareholders. Even without the use of DRM technology, content owners have the valid expectation that they would benefit

*IPTV Security: Protecting High-Value Digital Contents*   David Ramirez
© 2008 Alcatel-Lucent. All Rights Reserved

financially from their digital assets. IP rights include several areas such as copyright, patents, trademarks and design rights. The area that has a bigger impact on how contents are distributed via IPTV is copyright.

Copyright protects materials from being copied without permission and also extends to other activities such as making an adaptation, performing or showing the work in public, broadcasting and modifying the contents.

Copyright laws have been incorporated in most countries worldwide; some examples include an initial reference to the topic in the US constitution:

**1787: US Constitution**
Article I, Section 8, Clause 8 of the US Constitution, 'the Congress shall have power. . . to promote the progress of science and useful arts, by securing for limited time to authors and inventors the exclusive right to their respective writings and discoveries' [1].

One of the most relevant additions to the copyright legislation was the Digital Millennium Copyright Act (1998). President Clinton signed the Digital Millennium Copyright Act (DMCA) [2] into law on 28 October 1998 (P.L. 105-304). The law's five titles implemented the WIPO Internet treaties, established safe harbors for online service providers, allowed temporary copies of programs during computer maintenance and made miscellaneous amendments to the Copyright Act, including amendments that facilitated Internet broadcasting with incorporation of licenses.

Among the most controversial provisions of the DMCA is Section 1201. This section defines the prohibition of gaining unauthorized access to a work by circumventing a technological protection measure put in place by the copyright owner, where such protection measure otherwise effectively controls access to a copyrighted work. This prohibition on unauthorized access took effect two years after enactment of the DMCA. Over those two years, the Library of Congress conducted a rulemaking proceeding to determine appropriate exceptions to the prohibition. Additional rulemakings will occur every three years.

Since inception, DMCA has been used several times to stop initiatives to break security systems. Section 1201 has been invoked, for example, when the content scrambling system (CSS) for DVDs has been broken and subsequently when copy protection mechanisms for music and video have been breached. Most recently, the Blue-Ray encryption model was broken and DMCA was used again to limit the disclosure of key information on the Internet.

IPTV will be used to provide several different types of material, from live coverage of concerts and sports games to home movies and public interest broadcasts. Some types of content may not need IP rights enforcement (home movies, public broadcast), but, for most of the content, IP rights enforcement is a crucial element that needs to be maintained. Content owners are starting to ask for IP protection as part of normal licenses, and broadcasting agents must understand the risks and technologies involved.

It is clear that content owners are starting to create business models where a mix of different delivery mechanisms are used. Many movie studios obtain revenue from theatrical releases, DVDs, Internet and other media. Appropriate security mechanisms are required to continue supporting new media and new technologies for releasing content.

In general, IPTV has very high requirements of protection for content. There are a significant number of people interested in breaking the security mechanisms to obtain digital-quality contents, which could be then sold or distributed without paying royalties to the content owners. Appropriate mechanisms should be deployed to ensure compliance with the agreements signed with content owners.

Intellectual property protection within IPTV is supported by a number of technologies, including DRM and access technologies. DRM includes a series of mechanisms that allow content owners to control how digital assets are accessed by third parties; for example, subscribers or even IPTV service providers. From a business perspective, DRM allows for monetization of digital assets and a structured business model. Within this model, content owners can define rules that would be applied to content distribution.

The DRM and IP rights technology is implemented via software applications that must be secure to provide appropriate protection. Even if the encryption algorithms used are proven and accepted, the software may allow unauthorized access. This is particularly true for new DRM solutions that have been recently broken owing to unsafe implementations.

DRM has applicability to IPTV, B2B, B2C and C2C – in general, any exchange of digital assets or digital information that must be protected.

Figure 4.1 illustrates the conventional process flow for intellectual property protection. Business requirements regarding privacy, access to assets and fair use were articulated by a legal framework via patents, copyright and trademarks. Authors had the rightful expectation of benefiting from their intellectual creations and those rights were protected via a number of legal mechanisms including detection of infringing works, enforcement and litigation. This flow is valid for modern artistic works, but it can be translated and adapted to new technologies to ensure that the enforcement function and business requirements are translated faster into actions.

Originally, IP rights for artistic works were applied to music and literary works. At that time, most creative works were released either in printed form or a limited number of painted images. This environment was relatively easy to control. Printed or painted assets were easy to track, distribution of vast volumes was impossible and criminals could be tracked down.

In the late 1800s and early 1900s the general public could not turn on a radio and listen to music. The only forms of entertainment were live performances which were not practical for family environments at home. Public radio broadcasting started in the second decade of

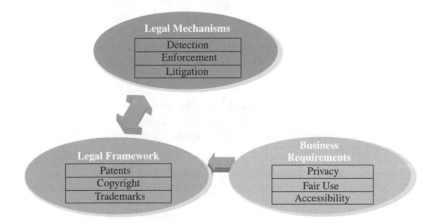

**Figure 4.1**  Intellectual property – flow

the 1900s. An alternative to live performances was the use of perforated piano rolls used to instruct the piano which notes to play; this was a new application of technology that caused very interesting effects on the legal plane.

As piano roll players became more widely used, the companies printing and selling sheet music experienced a reduction in sales (as probably did the piano teachers across the USA). This situation was presented to the US courts in the case between the White-Smith Music Publishing Company v. Apollo Company in 1908 [3]. The plaintiff, White-Smith Music Publishing, alleged that the defendant, Apollo (makers of piano roll players), had infringed the copyright on two songs published by the plaintiff ('Little Cotton Dolly' and 'Kentucky Babe', the copyright of which was assigned to White-Smith by Adam Geibel, the composer) by making piano rolls of them.

The ruling of the court presented their interpretation of what was considered a copy. Evidently, the court was thinking only in terms of paper and ink and did not consider new technologies.

The ruling quoted a number of decisions from other courts:

[Kennedy v. McTammany, 33 Fed. Rep. 584.] The decision was written by Judge Colt in the First Circuit; the case was subsequently brought to this court, where it was dismissed for failure to print the record. 145 U.S. 643. In that case the learned judge said:

'I cannot convince myself that these perforated sheets of paper are copies of sheet music within the meaning of the copyright law. They are not made to be addressed to the eye as sheet music, but they form a part of a machine. They are not designed to be used for such purposes as sheet music, nor do they in any sense occupy the same field as sheet music. They are a mechanical invention made for the sole purpose of performing tunes mechanically upon a musical instrument'.

[Stearn v. Rosey, 17 App. D.C. 562]. Again the matter was given careful consideration in the Court of Appeals of the District of Columbia in an opinion by Justice Shepard, in which that learned justice, speaking for the court, said:

'We cannot regard the reproduction, through the agency of a phonograph, of the sounds of musical instruments playing the music composed and published by the complainants, as the copy or publication of the same within the meaning of the act. The ordinary signification of the words "copying", "publishing", etc., cannot be stretched to include it.

'It is not pretended that the marking upon waxed cylinders can be made out by the eye or that they can be utilized in any other way than as parts of the mechanism of the phonograph.

'In these respects there would seem to be no substantial difference between them and the metal cylinder of the old and familiar music box, and this, though in use at and before the passage of the copyright act, has not been regarded as infringing upon the copyrights of authors and publishers'.

The court even included a reference to English courts, which showed how rulings at different courts tried to harmonize a consistent approach to copyright. In general, copyright tends to be equivalent or at least compatible across the world, something that is necessary to ensure business across borders, especially with digital assets that are created in one country and then broadcast in many different countries.

The question came before the English courts in Boosey v. Whight (1899, 1 Ch. 836; 80 L. T. R. 561), and it was there held that these perforated rolls did not infringe the English copyright act protecting sheets of music. Upon appeal Lindley, Master of the Rolls, used this pertinent language (1900, 1 Ch. 122; 81 L. T. R. 265):

'The plaintiffs are entitled to copyright in three sheets of music. What does this mean? It means that they have the exclusive right of printing or otherwise multiplying copies of those sheets of music, i.e., of the bars, notes, and other printed words and signs on these sheets. But the plaintiffs have no exclusive right to the production of the sounds indicated by or on those sheets of music; nor to the performance in private of the music indicated by such sheets; nor to any mechanism for the production of such sounds or music.

'The plaintiff's rights are not infringed except by an unauthorized copy of their sheets of music. We need not trouble ourselves about authority; no question turning on the meaning of that expression has to be considered in this case. The only question we have to consider is whether the defendants have copied the plaintiff's sheets of music.

'The defendants have taken those sheets of music and have prepared from them sheets of paper with perforations in them, and these perforated sheets, when put into and used with properly constructed machines or instruments, will produce or enable the machines or instruments to produce the music indicated on the plaintiff's sheets. In this sense the defendant's perforated rolls have been copies from the plaintiff's sheets'.

Following the principle of harmonizing with international agreements and law, and even if the USA was not part of the Berne convention of 1886, some of their conclusions seem to have been considered by the US congress and benefited the amendments to the 1870 Copyright Act enacted by the Fifty-first Congress on 3 March 1891. Section 13 of the act stated:

The International Copyright Act – Amendments to the 1870 Copyright Act [4] enacted by the Fifty-first Congress on March 3, 1891:

SEC. 13. That this act shall only apply to a citizen or subject of a foreign state or nation when such foreign state or nation permits to citizens of the United States of America the benefits of copyright on substantially the same basis as to its own citizens; and when such foreign state or nation is a party to an international agreement which provides for reciprocity in the granting of copyright, by the terms of which agreement the United States of America may, at its pleasure, become a party to such agreement. The existence of either of the conditions aforesaid shall be determined by the President of the United States by proclamation made from time to time as the purposes of this act may require.

The ruling on White-Smith v. Apollo and the lack of existing legislation on the subject caused new regulations to be created by the US congress. The text of the court decision made some references to the lack of copyright protection for conceptions:

The statute has not provided for the protection of the intellectual conception apart from the thing produced, however meritorious such conception may be, but has provided for the making and filing of a tangible thing, against the publication and duplication of which it is the purpose of the statute to protect the composer. . . . It may be true that the use of these perforated rolls, in the absence of statutory protection, enables the manufacturers thereof to enjoy the use of musical compositions for which they pay no value. But such considerations properly address themselves to the legislative, and not to the judicial, branch of the government. As the act of Congress now stands we believe it does not include these records as copies or publications of the copyrighted music involved in these cases.

The US congress enacted the Copyright Law of 1909, which created the concept of the mechanical license fees, and then these evolved into the performance royalties known today, paid to copyright owners when a piece of music is played in public.

A more recent case showing how technological changes have impacted the copyright laws is the Supreme Court decision in Sony Corporation of America v. Universal City Studios. The plaintiff (Universal) alleged that the defendant (Sony) produced equipment that could be used to infringe copyrights held by the plaintiff. The decision was against the Hollywood studios, and over time their business model has been adapted to deal better with this new technology and media. The lessons learned by the studios in the home video recorder world were then applied to the CD and DVD worlds, with more flexibility and lower prices. However, new technologies such as DRM were deployed on DVDs previously unavailable on video tapes.

Figure 4.2 illustrates a more fluid approach to intellectual property rights protection. The legal framework is translated into potential publishing policies. Based on the business requirements, some of these policies would be selected for a particular customer or market. The publishing policies would then be translated by the DRM functions into licenses that would have embedded portions of the enforcement mechanisms. For example, content would be encrypted until a valid key is presented.

The new technologies available allow content owners to use publishing rights policies and specify how content will be enjoyed. The legal framework is used and translated into publishing rights policies. This is articulated with the business requirements and market environment.

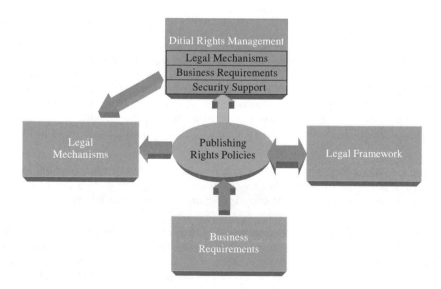

**Figure 4.2**   Intellectual property – new model

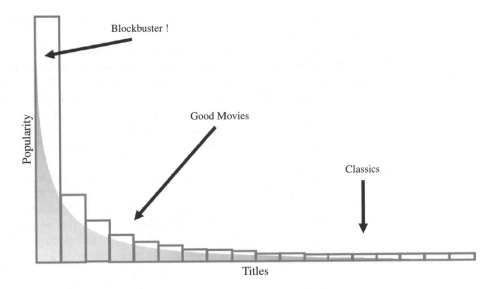

**Figure 4.3**   Long-tail effect

There is a simple concept that could be used to represent the appetite of subscribers for content – a decreasing curve that tends to be zero. This is illustrated by Figure 4.3. A few blockbuster movies are requested by a large number of subscribers, some recent good movies are requested by a fair number of subscribers and there are a large number of classic movies requested by a small number of subscribers. This concept is frequently applied to retail sales, where some few key products attract a lot of attention and a large number of products bring small sales. Revenue can be equivalent from both sources. IPTV service providers may end up with similar revenue from a few blockbuster titles and thousands of classic movies.

One example would be the possibility of renting access to videos. The legal framework allows for this type of access and there is a business requirement to increase the number of subscribers having access to digital contents, especially following the concept of the long tail, which suggests that the great majority of digital assets are of least interest to subscribers. Content owners would have a very limited number of blockbusters and a long list of classics and old movies. By providing rental access to the whole portfolio, a large number of subscribers will be interested in the service. The conditions of this access are defined using the publishing rights policies and deployed using DRM

Legal mechanisms support the operation of the service. For example, digital watermarks can be used to detect breaches to the copyrights and find the individual suspected of being responsible for the breach.

IPTV service providers must ensure that the service is operated within the acceptable legal market. This legal market has boundaries defined by the content owners and it allows for better consumer value, consumer protection against fraud, artist compensation and a sustainable business model for all parties involved. Market interactions will serve as catalysts to force a commercial entropy process between consumers and content owners. If prices are too high for consumers, then sales will be affected and additional piracy forms will be deployed. Fair and acceptable prices will facilitate the process of maintaining digital assets within the legal market. Legal enforcement is also an important mechanism as well as the technological tools available to protect contents.

Any digital asset entering the legal market must be watermarked to ensure identification. Additionally, fingerprinting functions must be applied to simplify the process of detecting unauthorized copies. Once assets have been disclosed within the legal market, DRM and other technologies must be deployed to avoid assets being released into the underground and being disclosed without authorization. These concepts are illustrated in Figure 4.4.

One of the key advantages of DRM and access control technologies is the flexibility. Content owners tend to have assets in different environments, for example music, movies and books. DRM can control access to all types of assets.

In a converged environment, where consumers are able to access digital contents using different media, and service providers are offering packages with two or more services simultaneously, the DRM licenses can be used to allow flexible access to assets by consumers.

Once properly deployed, service providers can allow subscribers to use a mobile phone or set top box to access TV channels, or laptops and mobile phones to access music. All access will be controlled using a single digital license issued to the subscriber. As illustrated in Figure 4.5, IPTV service providers must be prepared to offer a range of contents over a range of access technologies. Subscribers may expect continuous access to a content stream, starting to watch it on their mobile devices and then switching over to their TV set. An example could be subscribers watching movies from their mobile phones while in transit to their homes and then switching over to their TV sets when they arrive.

DRM is deployed to protect the contents end to end. From the moment the content enters the IPTV environment it should be encrypted to avoid unauthorized access. Security mechanisms at transport and at the home end should protect against unauthorized access attempts. The DRM and access control mechanisms create a security domain that encompasses almost all elements within the IPTV environment. The scope is much larger than the network security mechanisms provided to block unauthorized access by third parties.

This security domain also provides a clear view of the risk environment. If even a single set top box is compromised, then the whole domain is open. Intruders may be able to exploit

**Figure 4.4**  Options to protect information assets

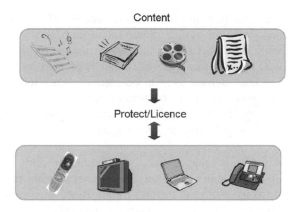

**Figure 4.5**   Types of consumer access

set top box vulnerabilities to extract digital content from the secure environment. Intruders attacking set top boxes will have significant amounts of time and will be able to work without interference from third parties. Clearly, this is much easier than trying to break into the head end to steal contents. Once vulnerabilities on the set top box have been discovered, intruders can replicate the attack on a large number of set top boxes and extract the content without being detected. Figure 4.6 illustrates the security domain. IPTV service providers must define clear security domains for content, deploying mechanisms that will protect contents at all endpoints.

Another risk is posed by the redistribution of contents. Intruders may be able to use set top boxes as distribution agents for content, releasing copies of content as it is being received.

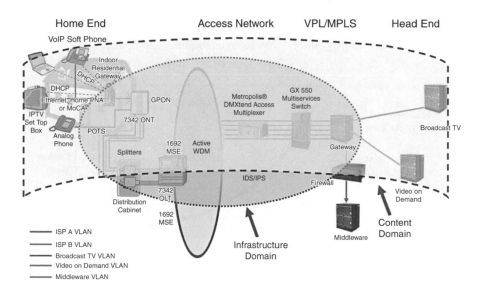

**Figure 4.6**   DRM domains

It is important to remember that subscribers need three specific elements to enable them access to digital contents: an encryption cipher (algorithm held by the set top box), a decryption key (issued by the DRM server and sent to the set top box) and a cipher text (encrypted movie). These three elements are more than is required to break encrypted files. Intruders would need at least two of them to be able to access the contents; usually the cipher and cipher text are used (as the key tends to be difficult to find). This situation brings to light an important concept. DRM and protection systems must be deployed matching the value and risk to the asset. A good way to visualize this fact is to look at the time value of digital assets.

As time passes, the intrinsic value of the asset diminishes. This is coherent with the concept of the long tail. On the thin part of the tail, fewer people are interested in the assets and it is not worthwhile to try and break the encryption. Assets also start on different points in the curve, with blockbusters starting with a high intrinsic value and close to the immediate timeframe; news and weather start with a low value and are already in the long-timeframe area. Intellectual property mechanisms must be deployed on the basis of the intrinsic value. Over time, assets will lose their intrinsic value, and a blockbuster from 1998 will not have a high value in 2008 as time has passed. Figure 4.7 illustrates this situation. All contents will lose value over time, and intruders will be less inclined to target particular assets. New releases may require stronger protections such as specific smart cards or tokens enabling access to the files. Subscribers may receive a pin number in the post, allowing them to unlock new releases.

The fact that subscribers are provided with the three elements required to open any content, and considering that content loses value over time, it is important to know that DRM and related technologies are only delaying unauthorized access and not blocking it. DRM systems will be broken over time, either because the software applications created to implement the DRM are weak and vulnerable or because the computing resources have evolved and keys from old algorithms can be guessed easily. Ideally, the DRM software implementation would

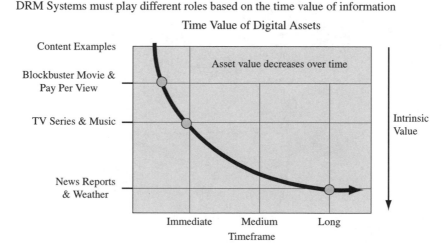

**Figure 4.7**  Time value of digital assets

be safe and AES and similar algorithms used within IPTV can last a few decades. By the time the algorithm is considered weak, the content will have lost its intrinsic value anyway.

Different implementations of DRM protecting digital video have been found to contain security vulnerabilities. More specifically, the protection mechanisms used by DVDs, Blue-Ray and HD-DVDs have been partially broken, allowing intruders to make copies of commercial discs. Additionally, a particular DRM implementation used by many content providers has been broken, and intruders are able to remove the DRM protection from digital files.

*Fair Use*

The term fair use is one of the concepts within the US intellectual property laws and allows for some limited use of copyright material without the specific authorization of content owners. In recent years, some content owners and content providers have decided to implement DRM restrictions to contents, and such restrictions have caused some consumers to lose access to purchased content. For example, some CDs may not work on car CD players and other portable CD players as they include DRM for music. Some DVDs are limited to specific regions. Music files downloaded from music portals tend to have restrictions and allow replay only on a specific handheld device. These restrictions have caused a consumer backlash, and many consumers are claiming for fair use to be allowed for legally purchased media. What consumers expect is files to work on any device without the need to preregister it. This is the flexibility to which they are accustomed from cassettes and CDs. However, for content owners this creates a high exposure to content theft and mass unlawful replication of contents.

## 4.2  Supporting Technology

The underlying technology for the protection of digital contents is based on two different cryptography methods.

### 4.2.1 Symmetric Key Cryptography

Secret key cryptography or symmetric key cryptography is a form of protecting data in which a secret key is used both to encrypt and decrypt the contents. The key then has to be shared using a different method or channel to avoid interception. Symmetric encryption algorithms include RSA, DES, 3DES, RC6, Twofish and Rijndael, the last ones being considered unbreakable by current available processing power owing to the large amount of time and resources required to break the keys using a brute-force approach.

Symmetric cryptography has several advantages including speed (considered at least 10 times faster than an asymmetric key), strength of algorithms and availability of implementations. The main problem with this technology is the key management/distribution and updates, as well as the scalability of the technology. Within an IPTV environment it would be insecure to deploy only symmetric keys for all the content as at some point the keys would have to be renewed and sent over the same network as the content. Intruders intercepting the communication would easily capture the keys and would be able to have access to encrypted content.

## 4.2.2 Asymmetric Key Cryptography

Public key or asymmetric key cryptography is a form of protecting data in which a pair of mathematically bound keys are derived from the same functions. At the same time the function reduces the possibility of deducting one from the other. Whitfield Diffie and Martin Hellman pioneered this concept in 1976.

Any user or system involved in the communication will have a pair of keys: one is called the private key and the other is the public key. If the user or system wants to receive encrypted information, the sender will be provided with a copy of the public key. This can be transferred using any technology. It is not dangerous to send the public key using insecure channels as the private key cannot be obtained on the basis of the public key (with appropriate key lengths the complexity is so high that it is considered impossible to guess the private key). Content encrypted with the public key of the recipient by the sender can only be decrypted using the private key from the sender.

Asymmetric key encryption has some clear advantages including key management and distribution and scalability of the infrastructure. Some disadvantages are that it is resource intensive compared with a symmetric key, which reduces the speed of the process.

Both symmetric and asymmetric encryption can be easily combined within an IPTV environment. Content can be encrypted using a symmetric key that is encrypted using the public key of recipients who in turn are the only ones allowed to open the key. The same network can be used to send the content and the key as it would be encrypted.

## 4.2.3 Hybrid Encryption

Hybrid encryption is based on a mix of symmetric and asymmetric encryption algorithms. This type of protection uses a random generated symmetric key that is subsequently encrypted with the public key of the recipient. This ensures that only the recipient will be able to open the encrypted contents and at the same time provides a faster process. The secure socket layer (SSL) is based on this principle. SSL is used across IPTV environments to protect the communications between set top boxes and the middleware server, as well as administrative access to many servers.

Hybrid encryption is deployed owing to the flexibility regarding key management and the speed provided by symmetric ciphers.

## 4.2.4 Hash – Digest

Hash or digest functions are one-way mathematical functions. These take a value as input and create a consistent output given the same input. If the input is modified even on a small percentage, then the output will be changed.

These algorithms are used to provide a reduced representation of the original text, and the resulting hash value can be used as a master version in order to detect changes to the original text. Checksums, hash and digest values are used within security functions to detect if a file has been altered. They can also be used to ensure messages are not modified. If a particular hash is encrypted with a private key, then recipients will be assured that the owner of the key was present at the moment the hash was created (thus a digital signature).

A secure hashing algorithm (SHA-1), message digest 5 (MD5) and others are used within IPTV to protect contents from modification in conjunction with encryption mechanisms.

## 4.2.5 Commonly Used Algorithms

*(i) Data Encryption Standard (DES)*
One of the most widely used encryption standards, promoted by the US government for federal and government communications, the algorithm was released in 1976 [5].

This cipher encrypts a fixed-size block of data. DES takes 64-bit chunks of data through 16 iterations of the algorithm or key. The 56-bit key length used by DES generates over 36 000 000 000 000 000 potential keys. The key space considered to be extremely safe in 1970 is now easily covered using modern processing power. An alternative to DES was the use of three iterations of DES (3DES) to increase the complexity of the process.

*(ii) International Data Encryption Algorithm (IDEA)*
A symmetric cipher. It is an eight-round cipher with a 64-bit block and 128 bit keys. The same algorithm is used for both encryption and decryption. It is considered to be very secure. The strength of the cipher is provided by mixing operations from different algebraic groups, which is resistant to both differential and linear cryptanalysis.

*(iii) Diffie Helman*
The principle behind this public key algorithm is the provision of a method to agree on keys over an insecure channel. The concepts behind the Diffie Helman algorithm facilitated the development of PKI and public key cryptography in general.

*(iv) Rivest Shamir Adelman (RSA)*
Probably one of the most widely used algorithms. It is based on the complexity of factoring large prime numbers [6]. The public and private keys are functions of a pair of large prime numbers. The basic decryption of the process for RSA is:

1. Take two large prime numbers, $P$ and $Q$.
2. Find their product $N = PQ$; $N$ is called the 'modulus'.
3. Choose a number, $E$, where $E < N$ and $E$ is relatively prime to $(P-1)(Q-1)$.
4. Find the inverse of $E$, called $D$, mod $(P-1)(Q-1)$, i.e. $ED = 1$ mod $(P-1)(Q-1)$.
5. $E$ and $D$ are called the public and private exponents respectively.
6. The public (encryption) key is the pair $(N, E)$ and the private (decryption) key is $D$.
7. The factors $P$ and $Q$ must be kept secret.

*(v) Advanced Encryption Standard (AES)*
The US government and National Institute of Sciences and Technology (NIST) started a process to replace DES as the recommended government standard for encryption. The process was undertaken in 1997 and ended with the selection Rijndael. The AES is a federal information processing standard (FIPS) FIPS-197 [7]. This standard specifies Rijndael as an FIPS-approved symmetric encryption algorithm that may be used by US government organizations to protect sensitive information.

The Rijndael is a block cipher, designed by Joan Daemen and Vincent Rijmen as a candidate algorithm for the AES. The cipher is based on a variable block length and key length. Currently specified is the usage of keys with a length of 128, 192 or 256 bits to encrypt blocks with lengths of 128, 192 or 256 bits, where all nine combinations of key length and block length are possible.

## 4.2.6 Public Key Infrastructure and ITU-T Recommendation X.509

Asymmetric key encryption is required as part of a solution to ensure nonrepudiation and additionally supports applications with symmetric key encryption to provide what is known as a public key infrastructure (PKI).

There are two relevant concepts that support the technology and are presented in ITU-T recommendation X.509:

### (i) Digital Signature

Digital signature is the process of taking a hash of a text and encrypting it with the private key of the sender. The encrypted hash is considered as the digital signature of the original text. Both the original text and the signed hash are sent to the interested party.

When the message arrives with the encrypted hash attached, the recipient repeats the hashing process, decrypts the attached file using the sender's public key and compares the two values which must be identical. This guarantees that the document was present at the sender's and that the sender confirmed the contents of the file by using his or her private key. This guarantees nonrepudiation of e-commerce transactions.

With public key encryption, text encrypted with the private key can be decrypted using the public key. This supports the process of digitally signing a document and allows senders to distribute their keys for verification to all interested parties.

Figure 4.8 shows how digital assets are digitally signed and then how the signature can be verified at reception.

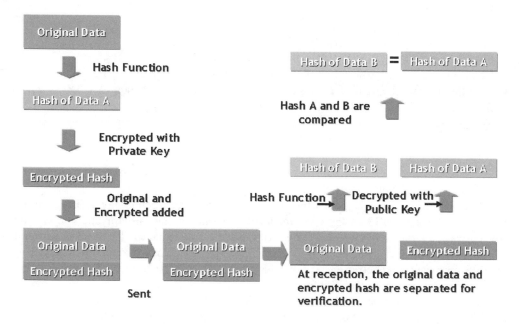

**Figure 4.8** Digital signature – example

*(ii) Digital Certificate*

Digital certificates are the equivalent of a passport. They can be used to prove one's identity. A digital certificate is issued by a trusted third party, containing a number of fields with information about a person's identity, including the person's public key. Digital certificates are stand-alone text files that can be exchanged by applications and presented by the user as part of the authentication process.

The trusted third party signing the certificate, also known as the certification authority (CA), digitally signs the content of the certificate to avoid manipulation of the data. Anyone who receives a digital certificate as part of an authentication process knows that the identity of the person is linked to his or her public key, and any data encrypted with that key can be retrieved only by the owner of the key and holder of the identity.

There are other elements within PKI, including:

- *The subordinate CA (Sub CA).* Used to relegate part of the authority and responsibility within the PKI environment. Very practical for geographically disperse deployments where thousands of users are at different locations. Sub CAs are formally authorized by a root CA who signs their digital certificate, including a field within the certificate that confirms the character of the Sub CA.
- *Registration authority (RA).* Used to register the information of the users and prepare the text that later will be signed by the CA or Sub CA.
- *Certificate revocation list.* This is a text file with the serial numbers of digital certificates that are no longer valid. The file is signed by the CA. This file is used by all elements within the PKI environment to check the validity of credentials presented by peers.
- *CRL distribution points.* URLs used to disclose the CRL to all interested parties. These points should be easily available to all members of the PKI environment.

Figure 4.9 illustrates the PKI framework, main components and information flow between them. Specifically, the dotted line between the root CA and sub CA denotes that under normal circumstances the root CA would be offline.

### *ITU-T Recommendation X.509*

In 1998 the International Telecommunications Union released the ITU-T recommendation X.509 [8]: *Information Technology – Open Systems Interconnection – the Directory: Public-key and Attribute Certificate Frameworks.* The document was intended for the following purpose:

This Recommendation/International Standard addresses some of the security requirements in the areas of authentication and other security services through the provision of a set of frameworks upon which full services can be based. Specifically, this Recommendation/International Standard defines frameworks for:

- – Public-key certificates;
- – Attribute certificates;
- – Authentication services.

The public-key certificate framework defined in this Recommendation/International Standard includes definition of the information objects for Public Key Infrastructure (PKI), including

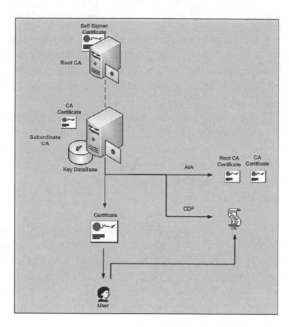

- Root CA
- Subordinate CA
- RA Services
- AIA Distribution Points
- CRL Distribution Points

**Figure 4.9** PKI framework

public-key certificates, and Certificate Revocation List (CRL). The attribute certificate framework includes definition of the information objects for Privilege Management Infrastructure (PMI), including attribute certificates, and Attribute Certificate Revocation List (ACRL). This Specification also provides the framework for issuing, managing, using and revoking certificates. An extensibility mechanism is included in the defined formats for both certificate types and for all revocation list schemes. This Recommendation/International Standard also includes a set of standard extensions for each, which is expected to be generally useful across a number of applications of PKI and PMI. The schema components, including object classes, attribute types and matching rules for storing PKI and PMI objects in the Directory, are included in this Recommendation/International Standard. Other elements of PKI and PMI, beyond these frameworks, such as key and certificate management protocols, operational protocols, additional certificate and CRL extensions, are expected to be defined by other standards bodies (e.g. ISO TC 68, IETF, etc.).

The X.509 also covers the Certificate Revocation List (CRL) implementation, and the Online Certificate Status Protocol (OCSP). These two functions ensure that the validity of the certificate is confirmed; this could be added to internal CRM or Financial applications to control content delivery to subscribers.

Figure 4.10 depicts how recommendation X.509 establishes the structure of a digital certificate. Extensions are a very useful field, allowing programmers to embed any relevant information including biometric data used to authenticate individuals using stand-alone devices not connected to a central database with records.

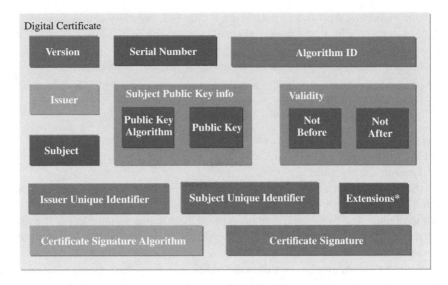

**Figure 4.10** ITU-T recommendation X.509v3 high-level structure

The fields have the following use:

- *Version.* This is used to determine the version of X.509 used for the structure of the certificate; v3 is the recommended version that is most widely used.
- *Serial number.* This simplifies the identification of a particular certificate. All certificates issued by a certification authority would have a specific serial number. Applications can refer to the serial number in order to identify a particular certificate. Additionally, the certificate revocation list is formed by the serial numbers of revoked certificates, added periodically and signed by the certification authority.
- *Algorithm ID.* This tells applications which algorithm was used to sign the certificate.
- *Issuer.* This would be the reference to the certification authority or sub CA that issued the certificate.
- *Subject.* Details of the subject represented on the digital certificate. Networking equipment, set top boxes and servers can own a digital certificate.
- *Subject public key.* This is a reference to the public key of the subject. This tends to be the 1024 or 2048 public key. Public keys can be disclosed without exposing the security of the private key. Digital certificates can be distributed publicly without significant risks.
- *Validity.* The certificate is not valid before a date and is not valid after a specific date.
- *Issuer and subject unique identifier.* Used to facilitate reuse of identities. Part of v2 and not widely used.
- *Extensions.* Included as part of X.509v3. Extensions allow for the addition of information to the certificate. This includes fields such as key usage and even biometric information.

- *Certificate signature.* Certificates are signed by the CA, basically using the private key of the CA to encrypt a hash of the digital certificate, as the public key is known, and it is easy to verify the signature.

Certificates have different applicability, one of the options for PKI use is authentication. The certificate can be used as follows:

- A sends a digital certificate to B for the purposes of authentication.
- B verifies if the validity fields are within an acceptable range. The certificate has not expired and is already valid.
- B verifies the information regarding the CA, specifically if the CA has been accepted/authorized before. Subjects can decide to accept a number of CAs.
- B uses the public key from the CA to decrypt the hash value and compares it with a fresh hash value computed on the text. If the two match, then the validity of the certificate is confirmed.
- B generates a random number and uses the public key from A to encrypt the number.
- A decrypts the random number using the private key and sends the decrypted number to B, confirming that A is in possession of the private key.

## 4.2.7 Operation of PKI

### Hierarchy Trees

The structure of PKI is based on trust relationships usually known as trees. The root is formed by the root certification authority (root CA) and branches are formed by the sub CAs. Leaves are formed by end-users, equipment or applications that use digital certificates and are not able to sign additional certificates.

The whole structure relies on the fact that the root CA is always protected and the private keys from the root CA cannot be manipulated or guessed. Trees are used to simplify the trust process. This way one leaf can trust another leaf and accept credentials as valid. At the same time, if an application has been configured to trust a number of root CAs, then all certificates coming from those particular CAs will be accepted as valid. Certificates coming from untrusted root CAs may require additional manual verification or will be blocked by applications.

### Operation

Applications are able to package the digital certificate of the subject as well as the certification chain. When received, the verification process will include checking that all certificates and signatures are valid and that none of them has been added to the certificate revocation list. The process starts with the subject that is being validated, going up on the hierarchy chain until verification that the root CA is trusted.

Set top boxes, servers and workstations would have some preinstalled root CAs as part of the web browser package. It is recommended that the list of root CAs be verified and all unknown CA be removed. This will ensure that only certificates coming from known CAs are accepted within the IPTV environment. If new CAs are accepted as part of broader business relationships, then root CAs can be added on demand.

## 4.2.8 Secure Socket Layer and Transport Layer Security

The secure socket layer (SSL) [9] was initially developed by Netscape in 1994 as a mechanism to support the development of e-commerce and safe communications on the Internet. This technology evolved, and SSL v2 was released in 1994. By that time some export restrictions on encryption technology were creating barriers to the type and length of keys that could be exported out of the USA. For many years, banks and other entities outside the USA were using weak key encryption, as their browsers and servers could not support longer key mechanisms. Even today, for compatibility reasons, many sites accept weak keys for SSL traffic.

Paul Kocher, Allan Freier and Phil Karlton started the development of the new version of SSL (SSLv3). It included new algorithms such as DSS, Diffie-Hellman (DH) and the National Security Agency's FORTEZZA.

In general, SSL acts as a hybrid encryption mechanism. When a browser wants to visit a particular website that is protected by SSL, the browser will create a random encryption key for the communication and will encrypt the random key with the public key from the server. The encrypted key is then sent to the server and, as the server is the only one capable of decrypting the key, the two machines will start encrypting all communications using the random key.

In 1997, the Internet Engineering Steering Group (IESG) [10] provided the evolution of the standard into transport layer security (TLS) which included the DSS algorithm for authentication, DH for key agreement and 3DES for encryption.

SSL and TLS are used for protecting HTTP traffic. Specifically, within IPTV they will protect the communications between set top boxes and the middleware. SSL channels allow clients and servers to send traffic instead of using the underlying TCP connections. The protocol specification *https* is used to indicate a request to open an SSL or TLS connection to the HTTP traffic (also referred to as HTTPS).

### Session Details

The initial TLS handshake process is based on the following process. The client and server will exchange initial messages to agree on algorithms, exchange random values and check for session resumption:

- Client and server will exchange the necessary cryptographic parameters to allow the client and server to agree on a pre-master secret.
- Both will exchange certificates and cryptographic information to allow the client and server to authenticate themselves.
- They will generate a master secret from the pre-master secret and exchanged random values.
- Provide security parameters to the record layer.
- Allow the client and server to verify that their peer has calculated the same security parameters and that the handshake occurred without tampering by an attacker.

Figure 4.11 illustrates the packet flow during the TLS session between set top boxes and the middleware server. Starting with information about the supported algorithms, the two then share information about the keys to be used and validation data.

**TLS Example**

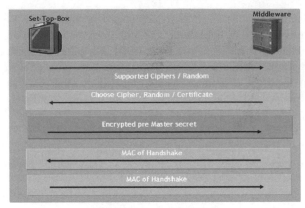

**Figure 4.11** TLS example

## 4.3 General Mechanisms for Content Protection

Within the IPTV market there are three primary types of technology used to protect video application intellectual property (IP) rights:

1. *Content protection systems (CPSs)*. Content is transmitted across networks in an encrypted form to help protect against theft or unauthorized access.
2. *Conditional access systems (CASs)*. These help to ensure that only authorized subscribers have access to the content and create a safeguard against theft of service.
3. *Digital rights management (DRM)*. This manages how the content is used by the subscriber on the basis of specific conditions set by the distribution contract.

The term DRM is accepted in the industry to include the CPS and it operates in conjunction with the middleware to provide CAS-related services. The DRM software supported by the DRM client on the set top box is able to encrypt contents from the source and issue licenses for access only by authorized subscribers; in this particular example, DRM acts as CPS, CAS and DRM. This chapter will present a basic description of the CPS and CAS services and will concentrate mostly on a DRM including the two other functions as well.

### 4.3.1 CPS

Content protection systems are used to ensure that content is only viewed by authorized subscribers. In both VOD and live IPTV, intruders can easily have access to the multicast and unicast streams. The purpose of the CPS is to ensure that unauthorized parties will not be able to decode the contents or redistribute the contents on the basis of the original stream.

To protect multicast sessions of IPTV, the head end will encrypt the contents after they have been encoded. There are several implementations and options for this protection. One would be to use a random symmetric key. This key would be used for either all content from the head end or content for a specific channel, or it could be changed during the day. This key would be handled at the application level and would reinforce the restrictions used by IGMP to allow only a group of hosts to have access to broadcast information sent.

The unicast situation for VOD is different because the head end will encrypt all the content with a random symmetric key only available for the paying subscriber. Then the key will be provided to the subscriber using the subscriber's public key to encrypt the symmetric key. Once the subscriber has received the symmetric key, he or she will be able to decrypt the content. Additional security measures include changing the symmetric key during transmission to increase the complexity of the process and reduce the chances of intruders finding or guessing the key.

Handling of the key management process, including how keys are exchanged between parties and change frequency, is an important security issue.

Within IPTV, CPS is one of the mechanisms more widely accepted by content owners for the protection of digital assets. It is also the most fundamental protection that they would require as it ensures that, once the content leaves the head end, it will be protected in transit until it reaches the subscriber and once there it will remain within some minimum boundaries restricting reproduction and redistribution (some of these functions are shared with DRM).

### Ismacryp
Within the realm of CPS there are some specific components that form part of the MPEG-4. The Internet Streaming Media Alliance (ISMA) has created Ismacryp, an end-to-end content protection model supported by MPEG-4. The approach of this standard is to encrypt and then packetize, providing end-to-end content protection.

## 4.3.2 CAS

Conditional access systems (CASs) are used by cable operators to control access to content. The CAS has evolved from simple frequency shifts and electronic noise to content encryption. Because telecom companies and carriers face different threats associated with IPTV, a variety of new technologies have been developed to protect data.

The CAS function is based on the same principles as CPS: once information has been encrypted, the system will ensure that session keys are only sent to those subscribers authorized to receive the content. In some cases, CAS can be deployed using access control lists only, without encrypting the content.

Internal verifications will take place to ensure that only valid subscribers can request VOD titles, and this includes checking account status.

## 4.3.3 DRM

Content owners realize that IPTV provides a great channel but is also a huge risk. The growth of peer-to-peer networks demonstrates that digital content can easily be traded on the Internet with little content owner control and without proper recognition of IP. Broad distribution of illegal copies would undermine the business model around digital media and would erode the revenues of content owners.

Technology providers initially implemented solutions that lacked strong DRM control, some using basic scrambling technologies or weak encryption mechanisms. Today, in spite of current baseline DRM controls, some vendors are still failing to address the issue of recording and replay – subscribers can store copies of DRM material and redistribute it on the web. This has prompted content owners to require content providers to implement strong DRM controls to reduce the risk of unauthorized access to content.

There are currently two basic options for content delivery using IPTV – video on demand (VOD) and broadcast. Each has its own particular security and DRM requirements:

- For VOD it is generally recommended that the content be segmented and encrypted using a symmetric key. The key can be changed several times in a typical movie to increase protection. Each set top box has the subscriber's private key, and the VOD server sends the encrypted content and the encrypted symmetric keys to the set top box for decryption and playback.
- Broadcast content follows a very similar process. Content is encrypted at the source with a symmetric key. The set top box sends a request for the current content key, and the content server sends an encrypted symmetric key to be used by the set top box to retrieve the contents.

Current industry trends indicate that content owners are demanding relatively strong encryption for their data, including implementation of the advanced encryption standard (AES). It can be anticipated that this will shift threats from the content to the endpoints and the transport layer.

Other DRM requirements include the stipulation that DRM-protected information should be unintelligible after leaving the source. It should only be decrypted after it has arrived at the destination. This involves changing the security architecture to include a key repository that is used as part of the encryption process.

DRM systems rely on XML interfaces to link the middleware with the entitlement engine within the DRM service and issue appropriate keys. In many cases the DRM will issue an XML signed file with the rights assigned to a particular subscriber.

DRM applications are required to support the particular codecs selected by the IPTV service provider to protect the contents (for example, H.264, MPEG-4 and MPEG-2). In order to provide acceptable protection for contents, the video stream should be encrypted using SHA-1 for hash and integrity verification and AES for encryption functions. Key lengths should be higher than 128 (ideally, 256 for AES).

The DRM environment may or may not be supported by the use of hardware repositories for encryption keys; for example, chips within the IP set top box or smart cards that can be inserted in the set top box. Smart cards have been used extensively by the video satellite providers. There have been some known cases where the security of these platforms has been broken.

For the encryption of contents, it is required that the DRM server be able to do encryption in real time. VOD in some cases can be encrypted after it has been received and then stored for future use. For broadcast content a similar approach can be taken and contents encrypted when they are received, also frequently changing the access key to broadcast channels.

There are different ways in which the middleware server and DRM server will interact. Some deployments will use the middleware as the only server accepting requests from set top boxes. This would include requests for DRM keys and licenses. In other cases the DRM server may be able to allow the DRM client to send requests directly from the set top box. The sample traffic is presented in Figure 4.12.

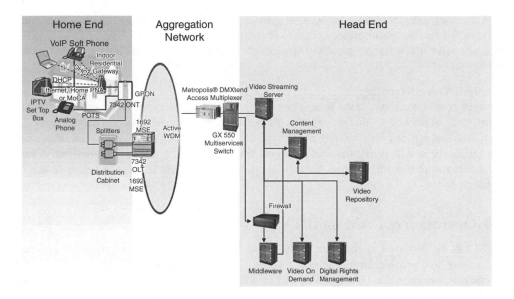

**Figure 4.12**   IPTV DRM and middleware interaction

Both VOD and broadcast channels will require encryption. In most cases this function will be done in real time. An additional function will be the exchange of encryption keys with set top boxes. These two functions will present a load on the DRM server; this component must be designed to support the appropriate number of requests.

The distribution of keys in broadcast IPTV can be done by sending the keys on a per-user basis. This method is more secure, although its scalability creates some problems. It is also possible to send the keys along with the content (ECMs) and send the ECM decryption key to all set top boxes, where it is saved in a secure storage (smart card). Then information on the user rights is sent as well on a one-to-one basis; this is also saved in the secure storage and is used by the terminal to decide locally if the content should or should not be decrypted.

### STB Client

The DRM client at the set top box will include capabilities to manage PKI interactions with the CA and other elements in the chain, as well as a repository for the private keys linked with the X.509v3 certificate assigned to the set top box. Key lengths will be determined by local regulations on encryption, processing power of the set top box and compatibility issues with other elements. The normal encryption found on this type of client is 1024 bit keys.

This acts as the client interface connecting with the DRM server to request keys and update certificates.

### DRM Encryption

For broadcast streams, the DRM must be capable of doing real-time encryption of contents. This would represent a load on the equipment and can be distributed using a number of servers dedicated for particular channels.

Some VOD contents are prepared beforehand and stored in encrypted form to be streamed at a later time. Video is encrypted using the compatible algorithms loaded on the set top box.

A standard encryption key for VOD is generated for all set top boxes. The key is changed frequently to avoid unauthorized access to contents. Some DRM implementations will change this key after a few hours; most DRM products will allow a personalized refresh time for the keys.

*Personal Video Recorder*
Encrypted content is sent to the set top box and the decryption key for the stream is encrypted using the public key of the set top box as per the digital certificate registered. Only that particular set top box will be able to decrypt the key to access the content, and, moreover, the DRM client will verify the content rights file linked to the movie to determine if there are limits to the number of times the movie can be watched and if there is a time restriction to block access to the movie.

## 4.4  Operation of DRM on IPTV

DRM can be deployed in different ways. Some vendors will opt for a very basic approach and others will provide an end-to-end solution. A typical DRM deployment would include interaction between the DRM server, the middleware, the video repository and the DRM clients at the set top box.

Figure 4.13 shows the different components of the DRM ecosystem. These include:

- the DRM system hosting the keys, licenses and entitlement information;
- the DRM clients requesting the keys;
- the middleware servers coordinating the information flow;
- the video repository hosting encrypted content and receiving encryption information from the DRM server.

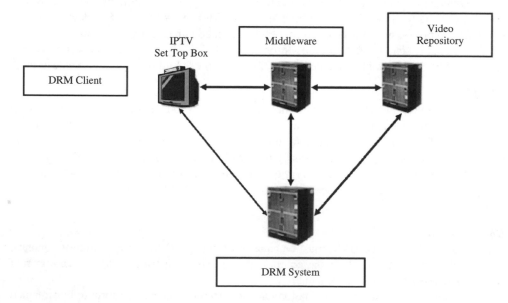

**Figure 4.13**  DRM components

## 4.4.1 DRM Applied to VOD

The video repository holds a number of encoded files ready to be streamed either to VOD or standard programming users. The process required to prepare and provide the VOD contents includes the following stages:

- encryption of original content;
- storage or broadcast of encrypted VOD content;
- key request from the set top box;
- authorization by the middleware and presentation of keys by the key management server;
- decryption of content by set top box.

### (i) Encryption of Original Content

Upon reception of the original content, either in a file or a stream, the encrypting function of the DRM system must require the encrypting key and algorithm information from the key management system. The key management system would have preconfigured the type of algorithms to be used and will generate a specific key for the VOD contents.

Once the key has been received, content will be encrypted using the selected algorithms and functions. In some cases the content will be divided in a way that, once encrypted, allows RTP and other VCR type functions.

Encrypted content is registered with the content management server and is sent to the video repository or the video streaming server.

### (ii) Storage or Broadcast of Encrypted VOD Content

Encrypted content can be stored on the video repository pending requests from subscribers. The content will be safe from unauthorized access as the keys are held on a separate system. Even when sent to a subscriber, the content remains under low exposure as the decrypting keys are provided using a separate VLAN; contents captured will not be accessible.

Stored content is registered with the content management service as well as with the middleware for the creation of the electronic program guide.

For streamed content, the stream is sent to the video streaming server and registered with the middleware and content management service.

### (iii) Key Request from Set Top Box

Once a subscriber has viewed the electronic program guide and selected a specific piece of VOD, a request will be sent to the middleware server regarding that particular piece of data. Subscribers have no information about the content yet, and they have to be validated before being allowed to have access to the contents.

### (iv) Authentication by the Middleware Server

The middleware will confirm the entitlement of the subscriber. Based on the business rules, packages and other authentication mechanisms, the middleware server will forward the request to the key management system within the DRM function for the key to be issued to the subscriber. Implementations of the DRM process will change from vendor to vendor. In most cases the public key of the set top box will be used to encrypt the content key, and this will be sent to the set top box. The content key tends to be the same for all subscribers, but each set top box will receive its own version.

*(v) Decryption of Contents*
The set top box will use either a file or a stream source and will apply the key received from the key management service within the IPTV service.

## 4.4.2 DRM Applied to Broadcast TV

Broadcast TV must also be protected to avoid unauthorized access to contents (access fraud) and breaches on copyright.

The broadcast process tends to have the following stages:

- encryption of broadcast content using session keys;
- content sent to the video streaming server;
- DRM server registers the encrypted content with the content manager and middleware;
- based on the EPG, the set top box requests keys for a stream from the key management server.

*(i) Encryption of Broadcast Content Using Session Keys*
The DRM system is responsible for encrypting the broadcast traffic, and it will request the session keys from the key management server. These keys will be provided to the set top box entitled to receive a particular stream. Keys can be the same for all channels and be renewed frequently, or each channel may need a specific key; the details will depend on the DRM vendor.

*(ii) Content Sent to the Video Streaming Server*
The content will be sent to the video streaming server for broadcast.

*(iii) DRM Server Registers the Encrypted Content with the Content Manager and Middleware*
The new encrypted content must be made available to subscribers via the middleware and electronic program guide. The DRM server will register this new content with the middleware so set top boxes can start to subscribe to a particular stream (join the specific VLAN).

*(iv) Based on the EPG, the Set Top Box Requests Keys for a Stream from the Key Management Server*
Once the subscriber has viewed the electronic program guide and has selected a particular stream that he or she is interested in receiving, the set top box will contact the key management server (some implementations will use the middleware to distribute keys) and will request the specific key required for the channel. Based on the duration of the key, this process may be repeated frequently during the day, or, in some cases, each time the subscriber changes channel a new key will be required.

## 4.4.3 Smart Cards and DRM

Smart cards are an inexpensive element that can be used to deploy security for environments with a large number of end-devices, such as set top boxes. The smart card acts as a repository for private keys and supports the process of authentication of set top boxes. The cryptographic processors embedded in the card allow for the protection and storage of keys. By using smart cards, set top box manufacturers can offer a second layer of protection for keys. Even if intruders are able to control the underlying operating system, it will be extremely difficult to extract the keys from the card. Additionally, cards can be replaced relatively easily by posting new ones to all subscribers and keeping the set top boxes.

The satellite and cable TV industries have used smart cards for quite some time, and, even with known cases of card fraud, they are still a viable mechanism to reduce the risks of unauthorized access and content fraud. DRM systems are currently being integrated with smart cards and private key storage to enable the encryption of content from the source and allow the creation of specific streams for each subscriber.

The smart card also facilitates the process of using different encryption keys during the broadcast process, thus increasing the complexity and resources required to break the security of the content.

Even after the smart card encryption has been breached, fraudsters have the added problem of distributing the cloned cards. The logistics related to this process create a barrier and reduce the impact of access fraud. (Without the cards, fraudsters need simply to distribute software modules that would enable unauthorized access to the contents. A similar situation occurs with regard to the 'content scramble system' (CSS) protection implemented for DVDs.)

Sections 4.4.4 and 4.4.5 have been extracted from the white paper 'DRM Convergence Analysis of Products and Standards' by Alvaro Villegas, IPTV solutions architect at Alcatel-Lucent [11].

## 4.4.4 Storage Protection

### 4.4.4.1 VCPS

The video content protection system is a standard defined by Philips for protecting video recordings transmitted with the broadcast flag and stored in a disc (recordable DVDs). It is included in the SCSI multimedia command set MMC-6 specification.

### 4.4.4.2 CPRM/CPPM (CPSA)

Intel, IBM, Matsushita and Toshiba have defined the content protection for recordable media and prerecorded media (within the overall content protection system architecture) in order to encrypt and control the usage of digital assets when recorded on any kind of media (recordable DVDs, flash memory, SD cards, etc).

## 4.4.5 Open DRMs

The term 'open' is used in this context to point out that the DRM is based on a published standard, so anyone can build a compliant client or server. Notice this does not mean that the usage of the standard is free of charge.

### 4.4.5.1 SDMI

In 1998 a consortium of around 200 companies was built with the common purpose of designing new technology for protecting the usage of digital music.

The DRM standards published by this Secure Digital Music Initiative included a watermarking algorithm that was submitted to the public community in September 2000 (SDMI challenge). Unfortunately for them, the scheme was deeply flawed and was therefore deemed unusable. The group has been inactive for several years.

### 4.4.5.2 OMA DRM

The Open Mobile Alliance is the answer of the mobile companies to the interoperability challenge in all technical aspects, which include, of course, the DRM. The first version of OMA DRM (1.0) includes some basic functionality (forward lock for ring tones and other downloaded items, mainly) and has been integrated in the vast majority of the mobile phones present in the market. The evolution of that standard is OMA DRM 2.0, which includes a much more complex architecture with a complete suite of tools and rules of usage that enable sophisticated business models for downloaded or streamed content.

The technology behind OMA DRM 2.0 is based on a rights expression language (open digital rights language, ODRL) and, in spite of its original target of being royalty free, is being claimed as falling in the ContentGuard patent pool. In fact, the main obstacle that prevents OMA DRM from becoming the universal DRM for the mobile world is the IPR claims that Contentguard, Intertrust, Matsushita, Philips and Sony (none of them members of OMA) have made on the OMA DRM design. MPEG-LA even published the expected license to be paid by any implementation, and it happened to be extremely high (and unacceptable for the mobile operators).

In the pure broadcast scenario, OMA has also published a standard called OMABCAST (with two profiles, based on device or smart cards), which is currently competing with DVB-CBMS (open security framework) to become the de facto standard for DVB-H.

Many providers already offer a commercial implementation of OMA DRM: BeepScience, CoreMedia, Discretix, Irdeto, Mutable, NDS, SafeNet, Philips and Viaccess.

### 4.4.5.3 DMP

The Digital Media Project is a nonprofit association registered in Geneva, Switzerland. DMP promotes the development, deployment and use of digital media that safeguards and balances the rights and commercial interests of creators, consumers and value-chain players, within the provision of product and services to the endpoints of the value chain and to other value-chain users.

The first specification of DMP (IDP-1) is focused on portable devices that depend on others (e.g. a PC) to get their content. The model is based on MPEG21 and other standards: XML content representation, rights expression language (REL), AES encryption, X.509 certificates, etc. DMP content information (DCI) can be packaged using either a file mechanism (DMP content format, DCF) or as a stream: DMP content broadcast (DCB) for an MPEG-2 transport stream or DMP content stream (DCS) for IP delivery.

The second spec (IDP-2), released in May 2006, addresses the DRM interoperability through the definition of a mandatory core set of DRM primitives and the optional inclusion of DRM tools (downloaded from a server or embedded in the content itself) that can expand the functionality when required.

Authentication in DMP is done on devices (IDP-1) and also on users (IDP-2).

Another interesting feature of the DMP is the analysis of the 'traditional rights and usages', a compilation of the DRM-related functionality and rules from the predigital era, with which end-users are quite familiar and which should therefore be taken into account in any new DRM scheme to ensure a nondisruptive experience and hence a good acceptance.

Although technically very attractive, the main risk of this initiative comes from the limited presence of content owners or big companies in the group.

### 4.4.5.4 MPEG21

This MPEG specification defines an open framework for multimedia application. It addresses the DRM functionality in two ways:

- MPEG rights expression language (REL), derived from ContentGuard's XrML;
- rights data dictionary (RDD): specification of atomic components in rights-related operations.

### 4.4.5.5 MPEG21 REL Data Model

MPEG-REL does not seem to be very successful as a stand-alone solution, although it has been incorporated in DMP. Microsoft uses its own flavor of XrML, and OMA uses its own open digital rights language (ODRL). RDD is not too successful either; other approaches such as Coral seem to be better accepted.

### 4.4.5.6 DVB-CPCM

DVB enters the DRM arena with its specification about content protection and copy management (CPCM). Oriented to the broadcast market, it works by including a set of flags in the content that describe how it should be used once it arrives at the subscriber's home. The flags cover five areas of control: copy and movement (copy once, no copy, copy many, etc.), consumption (watch once, 24 hour rental, number of concurrent playouts, etc.), propagation (authorized domain concept), output (digital and analog) and ancillary. All CPCM-enabled devices should behave appropriately depending on the flag value. A scrambling algorithm is included in the specification.

This standardization group is mostly led by content owners (MPAA, Disney, Warner, etc.) and tries to get a common position among many players (Microsoft, Nokia, NDS, Thomson, etc.) that in most cases have their own solution, so the advance has been quite slow and does not seem to be very active since the first publication of the BlueBook in November 2005. The full specification and implementation guidelines are due to be finished in 2007 and will then be submitted to ETSI for standardization.

### 4.4.5.7 DVB-CBMS

The ad hoc group 'Convergence of Broadcast and Mobile Services' of DVB has prepared a set of specifications and guidelines for the provision of broadcast TV services on mobile, which includes a conditional access framework for DVB-H that is based on the experience from the pure broadcast world. This standard is called the 'open security framework' and describes an architecture similar to the classical DVB Simulcrypt (including the ECM/EMM scheme), but with some additional parts (in the terminal architecture, mainly) that in the pure broadcast version were left open and resulted in independent, incompatible implementations by different CAS vendors.

A possible advantage of this solution when compared with the OMA equivalent standards is that it does not include any IPR that can be claimed by ContentGuard or Intertrust; the security details that might be patent aware lie in the proprietary part and are therefore handled by the companies providing the actual implementation (which, not surprisingly, will be the classical CAS vendors).

### 4.4.5.8 PERM

This working group of the IETF was created to define a security framework for content protection, including an authentication mechanism, a rights language and a strong encryption mechanism. PERM stands for Protected Entertainment Rights Management.

A draft was published in 2004, but there has been almost no activity in the last years in this group. It seems that the IETF itself was not really interested in this initiative (it was pushed externally).

### 4.4.5.9 DCAS

Standing for Downloadable CAS, this initiative from CableLabsIt is supposed to be a cheaper alternative to the CableCARD standard, which defined the implementation of the full CAS client in a detachable module (PCMCIA), in a way somehow similar to the DVB 'Multicrypt' architecture (common interface).

The DCAS approach is based on a standardized secure hardware (a customized ASIC chipset) on which the actual CAS/DRM implementation can be downloaded. This way, the operator can still get the higher security of proprietary solutions but with a very easy way to swap to a different vendor in case of necessity (piracy or simply commercial reasons). This swapping is in fact one of the hardest tasks and therefore the weakest points for the operator in a traditional proprietary CAS solution.

### 4.4.5.10 DReaM

DReaM stands for 'DRM everywhere available'. It is an initiative from Sun, started from their contribution to the EURESCOM Opera project, with the intention of creating an interoperable, open-source, patent-free, royalty-free DRM. A royalty-free codec is also being developed. The group is willing to work with like-minded communities.

Considered naive by some people, it seems to have a much narrower scope now. It has two parts:

- DReaM/CAS: open-source, royalty-free CAS for IPTV;
- DReaM/MMI ('Mother May I') is the DRM attempt, but it seems to suffer the same patent issues as the rest of the initiatives.

One of the interesting aspects of this project is that licenses are issued to individuals as opposed to devices (this is also known as personal rights management). The DRM is independent of content type, and it addresses customer requirements for multisystem interoperability.

### 4.4.5.11 OpenIPMP

Built by ObjectLabs, and promoted by ContentGuard, this published standard uses XrML (through MPEG-REL). The source code is available as a SourceForge project. Not much activity has been detected since 2003, although a new release was announced in July 2006.

It supports MPEG-IPMP for MPEG2 and MPEG4, OMA 2.0 and ISMA encryption and signaling.

### 4.4.5.12 OpenCA

This CAS specification is being proposed in the FLO forum, the consortium that is promoting the MediaFLO technology (a competitor of DVB-H) developed by Qualcomm in the USA, and now being pushed as a standard in the Telecommunications Industry Association (TIA).

The OpenCA architecture is conceptually very similar to the open security framework currently being proposed in the DVB-CBMS forum, and backed by some traditional broadcast CAS player (NDS, Nagra). These CAS providers are also supporting the OpenCA initiative for MediaFLO platforms.

### 4.4.5.13 PachyDRM

This term (a pun on 'pachyderm') refers to an originally proprietary DRM developed by Melodeo (a US company offering technology and also an online multimedia service for mobile), who in June 2006 decided to promote it as an open specification. They have also published the source code of their own implementation, which can be used and modified under specific licensing terms (in spite of what its promoters claim, it cannot be considered an open source initiative).

This DRM is oriented to the mobile environment, and, given that it is published as a standard, it has the same advantages and drawbacks as, for example, OMA DRM, including the IPR issues. It has been approved by major content owners and has been adopted (when it was still purely proprietary) by several relevant mobile operators worldwide.

## 4.4.6 Interoperability Proposals

### 4.4.6.1 Coral

This consortium presents one of the most active and promising initiatives for achieving a real DRM interoperability. Funded in 2004 by Intertrust, Sony, Philips, HP, Samsung, Matsushita and Fox (notice the absence of Microsoft and Apple), Coral aims to provide an ecosystem for content access that is independent of DRM, device and media format while at the same time respecting the right holder's rights. With a vast number of DRM, consumers might find it difficult and annoying to deal with different types of DRM. Making sure that consumers have a positive experience by making DRM invisible without defining a new DRM is the goal of Coral.

Beside its founders, Coral has quite a few interesting members such as Cisco, Motorola, etc., and also content providers such as Sony BMG, Time Warner, EMI, Warner Bros, Universal, etc.

Coral defines an ecosystem in which the Coral interoperability framework (core) ensures interoperability between a limited number of DRM systems. Coral plans to develop at least one ecosystem. It also ensures that there are no major changes to the DRM systems themselves and thus is a very good candidate as an interoperability solution.

The interoperability proposal is service based: independent, incompatible DRMs can interoperate through a translation service (including rescrambling when required) provided by an interoperability operator.

The overall architecture on which the Coral DRM proposal is based was published by Intertrust in 2004 under the name NEMO (Networked Environment for Media Orchestration).

A demonstration of the Coral platform with Microsoft DRM was presented in August 2006.

### 4.4.6.2 DMP

The Digital Media Project (see Section 3.5.3 for details) does not provide a universal 'DRM standard' capable of providing interoperability between users in different value chains or across different value chains. It provides instead some specifications defining a set of primitive functions that are derived from use cases, and examples of how value chains serving specific goals can be set up using standard tools. DMP hopes to build standardized tools, forming the so-called 'interoperable DRM platform' (IDP).

DMP is very active but lacks the clout of content providers, vendors and equipment makers that Coral or OMA boasts.

### 4.4.6.3 SmartRight

Promoted by a consortium of companies (Thomson, Nagra, Gemplus, Axalto, ST, Pioneer, Micronas, SCM, Panasonic), this DRM interoperability proposal can work with existing CA/DRMs through converter modules. It is based on smart cards, and no return channel is required for the operation.

The architecture is based on the concept of personal private network (PPN), a set of devices sharing access rights for content. The different devices at home may work as access gateways (content arriving from the outside), presentation endpoints, storage or exporting terminals. Once received in the PPN, the keys for accessing the content are reprocessed (re-encrypted) by the local smart cards using local keys. Key management in the PPN is fully managed by the smart card (including the addition or renewal of new members in the network).

Depending on the original DRM that was used for protecting the content that arrives at the PPN, a rescrambling may be needed in the access device. This is done transparently for the user. The SmartRight scrambling algorithm is DVB-CSA or TripleDES.

The copy protection provided by this protocol includes three usage states: copy free, private copy (the typical 'copy once') and view only (the classical 'copy never').

### 4.4.6.4 SVP

SVP defines an open content protection model that provides interoperability for consumers and devices. Content and its license (content keys and rights) are always inside secure chip silicon. There are no global secrets in SVP and therefore only a single device can be compromised, thus avoiding system-wide hacks. The content is encrypted by unique device key or domain key. When content goes from device A to C through B, unlike pipe protection, only A and C know the encryption keys. A higher-level view of the system is given below:

- CA/DRM business model enforcement;
- SW chip driver;
- TRS-SW enforcement (for domain, proximity, B-flag, private extensions);
- CA/DRM/FTA agent;
- usage model definition and billing;
- HW chip requirements, e.g. personalization, compliance and robustness;
- core security functions: certificate handling, SAC, CryptoTools, time, key management;
- content license (usage rules and content keys), export content control, revocation;
- content processing: content de/scrambling, content decoding;
- secure boot loader in HW.

### 4.4.6.5 OpenCP

In order to implement this architecture, SVP standardizes a set of DRM-specific features and requirements for the main chipset of the customer device. The specification has been implemented by the major chipset manufacturers, and can now be ordered as a standard product by, for example, STB providers. However, its success has been quite limited owing to the fact that its promoter (NDS) is a key player in the CAS/DRM arena, and competitors are reluctant to pay royalties to NDS for the SVP licensing.

### 4.4.6.6 OMArlin

This specification was announced in March 2007 by the Marlin Developer Community (MDC) as an attempt to link OMA DRM 2.0 and Marlin DRM standards:

- Devices compliant with the new spec will be able to access OMA or Marlin content transparently.
- A new common format, fully OMA and Marlin compliant, enables operators to prepare content that can be accessed in both OMA and Marlin terminals.

This effort has been led by the Marlin promoters (Itertrust, Philips, Sony, Matsushita and Samsung).

## 4.5 Watermarking and Fingerprinting

### 4.5.1 History

People have played with hiding information for centuries. Part of military and political history has always been linked with the need to hide information from opposing forces. As reported by the Greek historian Herodotus, messages were hidden in wood tablets that were then covered with wax to hide the message. Slaves had messages tattooed on their shaved heads and, once their hair had grown and covered the message, were sent to the destination.

As warfare evolved, so did the techniques to hide messages and information. The Spartans developed the skytale, a staff of wood of specific diameter that was used to wrap a leather belt or papyrus; the message was written along the length of the wrapped belt, and, once unwrapped, the meaning was lost until the original staff was used. Emperor Julius Caesar is credited with the invention of a cipher that was based on replacing letters of the alphabet with the corresponding letters three spaces forward.

A typical way to hide messages was the use of secret inks (lemon and milk was a favorite when I was a boy scout). This method uses a blank piece of paper. The secret message is written using a secret ink that can be seen on the paper surface before it dries; then a fake message can be written between the hidden lines using black ink. Once the recipient has the paper, he or she can use a heat source to bring out the secret ink.

During the Middle Ages and Renaissance there was a significant development of one-time pads and substitution ciphers. Famous writers such as Edgar Alan Poe are credited with a few examples of that type of writing.

Historically, the initial approach to protecting messages was ensuring that the enemy was not aware of the existence of the message in the first place. As techniques evolved, people started to trust their underlying protection mechanisms and then started to exchange information over open channels.

A recent example was the enigma machine used by the German forces during World War II, a complex piece of engineering that used a number of discs to encipher a message. However, also during WWII some spies used microdots to exchange messages with handlers; these messages had the size of a dot from a typewriter and could be added to a normal paper with any text.

In current times there are two main ways to protect information: cryptography and steganography. Cryptography is used to protect the information when sent over open channels and exposed to interception. Steganography is used to hide the fact that information is being exchanged. The term 'steganography' has Greek roots: 'steganos' (hidden) and 'graphia' (writing). With steganography, attackers do not know that there is secret code on the carrier image, and they do not have information on where to find potentially secret information.

There are different ways of protecting data. Encryption is used when there is an expectation that content will be protected and the message is likely to be intercepted and analyzed. Some typical encryption models include transposition and substitution ciphers.

When the message is unlikely to be intercepted, or there is no expectation that the message will be protected, a covert protection mechanism can be used such as steganography. Figure 4.14 shows the relationships between the different mechanisms.

Steganography has varied applications in modern life. It can be used to exchange communications without alerting the transporting parties of the existence of a message. There are some applications related to privacy and limitation of liability. This technology can also be deployed to protect the exchange of information and facilitate access to censored materials.

The process of embedding information on a carrier media (text, music, video, images) starts with the selection of a cover image. A key (k) and secret message (m) are added using an encoder. The resulting stego-object (s) would be equivalent to the original and, even with attempted verifications, the message would remain secret. The new stego-object (s) is sent over any channel, including exposed/insecure channels. Once received, the stego-object (s) is passed through the decoder (the decoder must be compatible with the encoder), and, with the addition of the key, the message is extracted. Figure 4.15 illustrates the process.

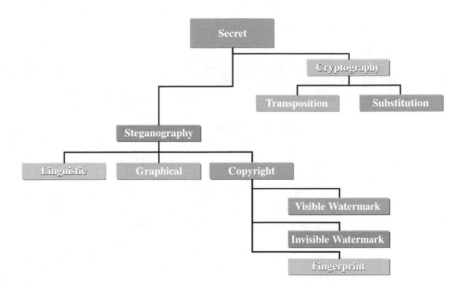

**Figure 4.14**  Mechanisms for protecting information

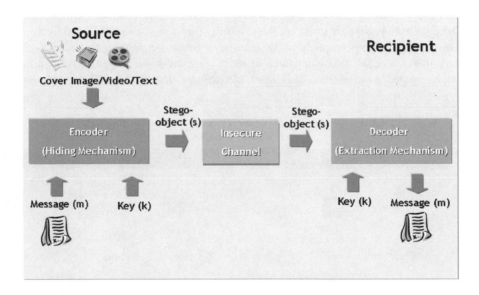

**Figure 4.15**  Basic steganography process

Most relevant to IPTV are the techniques of invisible watermark and fingerprint. These are used to mark content with information related to the subscriber who is receiving the content. For IPTV implementations, steganography and cryptography can be used to encrypt the codes used to identify subscribers and avoid detection.

## 4.5.2 Steganography Techniques

### Substitution Technique

Substitute parts of the carrier object. These parts can be considered redundant or unnecessary and will not affect the overall perception by users or systems.

A few examples of this technique are the least significant bit (LSB) substitution and the MPEG-2 and MPEG-4 4 carrier.

The LSB method can be explained better with an example of how to use it on an image. This method works on images using bytes to represent colors, specifically following the structure of values in 24 bits per pixel (bpp). The color of each pixel is represented by three integers between 0 and 255, representing the intensity of red, green and blue. For image standards that use a palette, this method does not apply. Based on the full range of colors, a pixel can take any of 16 million options. The human eye is not able to detect minute changes in the colors of pixels, and this is used to support the LSB method (a pixel can take any of $2e^{\wedge}(8*3) = 16\,777.216$ values).

Each pixel is represented by a string of 24 bits. These 24 bits are split into three blocks of eight bits used to represent red, green and blue. If the last bit of each color is changed, then the color of the pixel will suffer a minuscule change; a variation of three units within 16 million.

One example, as illustrated in Figure 4.16, could be a pixel with gray color. In this example the RGB values are 128, 128, 128, In binary this number would be represented as 10000000. The last bit of this string can be changed from 0 to 1, resulting in 10000001 which represents 129. Once the three LSBs are changed, the resulting gray would be represented by 129, 129, 129. This change is not perceptible to the naked eye. Some steganography applications may use more than one LSB. Changing three or more may cause the image to be significantly altered and tagged as suspicious. It is important to note that, within IPTV, steganography

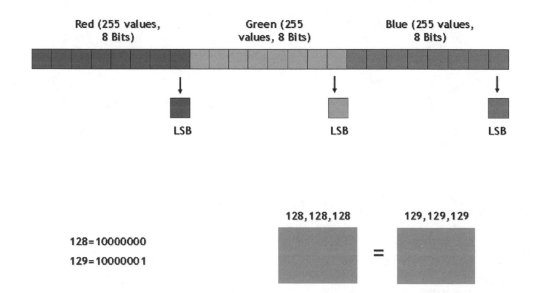

**Figure 4.16**  Least significant bit applied to RGB

would be used to mark contents. These marks are small and will never require more than one LSB. Additionally, the original content will not be available to intruders, and hence detection and removal are very difficult.

### Transform Domain Technique

This technique is based on the concept of using the transform space (frequency domain) of an image as the carrier [12]. A common application is the discrete cosine transform (DCT) domain. The carrier image is divided into $8 \times 8$ matrices, where each matrix is used to encode one message bit. Within the DCT, matrices are chosen in a pseudorandom manner, and the size of two predefined DCT coefficients is modulated using the message bit.

### Spread Spectrum Technique

This concept is based on spread spectrum communication, where a signal is broadcast embedding information in the signal. Messages are spread over a wide frequency. Different bands are available to split the message, providing additional reliability.

The principle of spread spectrum steganography is based on concepts from radar technology. Radio signals tend to have a pulse info surrounded by radio noise in the rest of the spectrum. This narrow band is changed over time, following a pattern that could be defined as the communication key. For an unsuspecting party monitoring the communications, this narrow/split-second communication may seem to be noise. Any attempt to intercept the rest of the information would result in a wild chase across the whole spectrum as the short bursts of data are never on the same frequency and never on surrounding frequencies.

## 4.5.3 Watermarking and Fingerprinting Principles

Watermarking is the process of imperceptibly modifying a carrier to embed a message, with the message carrying information about the cover. Information can include ownership, access rights, serial numbers or the last equipment that displayed a video.

Historically, the process of using modified rolls to modify the surface of paper during the manufacturing process has roots in Italy from around the thirteenth century. The main reasons behind watermarks were the need to identify the manufacturer of a quality product and later to avoid forgery of valuable documents. Many bonds and notes use watermarks as a counterfeiting protection.

In 1954, Emil Hembrooke filed a patent to identify music works, and this is referred to as one of the first patents related to marking contents. Later, in 1990, Komatsu and Tominaga published the first references to digital watermarking [13].

Watermarking is an application of steganography, and within IPTV it is used to embed on the video stream information about the content owner and rights and to fingerprint the last equipment that displayed the content (set top box or computer). Both watermarks and fingerprints need high levels of resilience against modifications of the carrier video stream. They must survive encoding and other alterations. In general, these techniques have three main applications.

### (i) Copyright Protection

From a legal point of view, this is a secret mark made by the content owner. The mark includes information about who owns the content and is used to avoid third parties claiming ownership of a particular asset.

## (ii) Rights Management

A watermark is added to define digital rights of that particular copy. For example, the content may have a flag authorizing only playback on digital sets or not allowing copies to be made.

## (iii) Content Validation

Watermarks can be embedded to identify any changes to the stream. If the content is modified or altered by third parties, then these changes will be easily detected by the changes in the watermark.

### Transaction Tracking

A watermark is generated by the last element on the transmission chain and it includes information about the end-user. This may include the customer code and IPTV service provider code or similar information that can be used to reconstruct the path followed by the particular copy of a digital asset.

This fingerprinting technique (similar to the trails left by criminals while present at the crime scene) is used as a forensic tool to help track leaks in the IPTV environment. An extension of this technique may include content owners marking video contents before they are delivered to the IPTV service provider and stating who the distributor and IPTV service provider are. If content is recovered from underground sites or black-market DVD copies, it would be viable to ascertain who lost the asset.

Watermarking and fingerprinting must be deployed with an understanding of the balance and entropy relation of the three main security characteristics of the steganography techniques. Any steganographic mark would have to balance resilience, stealth and capacity. A very large message will be difficult to conceal and will be damaged once the content has been encoded, resized or cropped. Equally, an almost invisible message, very hard to detect, will be small and may present some problems when the carrier has been modified, cropped or resized.

As depicted in Figure 4.17, IPTV service providers could achieve high levels of resilience at a cost to the secrecy of the mark. Similarly, a highly secret mark may have low levels of resilience. The type of information stored is relatively small, and hence capacity may not be an area of concern. For applications outside watermarking and fingerprinting, capacity would have to be considered.

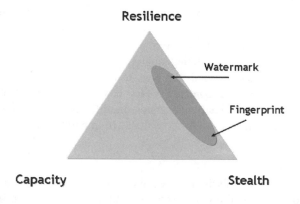

**Figure 4.17** Entropy within watermarking

Within the scope of these two techniques, fingerprints tend to be small and stealthy, and watermarks tend to be small and resilient. IPTV video streams are large in size, and embedding these two types of mark is relatively easy. Finding these marks or modifying the carrier to eliminate the marks is a very difficult process.

## 4.5.4 Typical Attacks

There are a number of attacks that can be used to detect, retrieve or eliminate digital marks embedded in content carriers. Some of the most commonly used are as follows.

### (i) Lossy Compression
This technique is a compression technique in which some amount of data is lost. During the process, the algorithm will attempt to eliminate or make redundant unnecessary information. MPEG codecs are based on this principle (also JPEG, H.26x and others)

Some examples of lossy compression are:

- joint pyramid encoding;
- joint difference pyramid encoding;
- rate-minimising pyramid encoding;
- DCT coding of combined image blocks;
- pixel-based disparity map encoding;
- block-based disparity map encoding;
- pioneering block-based predictive encoding;
- compression using subsampling and transform coding.

These algorithms will discharge some of the fields used by the steganographic algorithm during the watermarking process. The main risk from lossy compression applied to a marked asset is that it may lose some (or all) of the fields selected to hide the mark. Within IPTV, most watermarking techniques use the fact that marks are very small and a high number of copies can be stored. Some marks may be lost, but this method is highly resilient as only one mark is needed to identify the content.

### (ii) Geometric Transforms
Translation, rotation and scaling can be used to make alterations to a movie or image. These are based on the changes to the coordinates or perspective on an image and may affect the embedded information.

### (iii) Data Reduction
Cropping, resizing, clipping and removing unwanted areas of the stream may have an impact on the stored marks.

### (iv) Jitter
Jitter is used to define the deviation of some aspects on a digital signal. Deviations can be in terms of amplitude, phase and width of the signal. Adding jitter may break the watermark or embedded digital fingerprint.

## (v) Enhancement

Using software tools to change the contrast, sharpening the image or filtering out colors or shades, may affect the watermark. Some watermark algorithms are specifically designed to withstand this type of attack.

## (vi) Noise

Adding noise to a carrier signal will replace some of the fields used to store the fingerprint or watermark and render the information unusable. For IPTV, the size and number of copies of the fingerprint protect the information and support higher resiliency.

## (vii) Filtering

Filters can be applied to modify images, signals and video streams. A high-pass filter allows high frequencies and reduces frequencies lower than the cut-off frequency. A low-pass filter allows low frequencies and reduces frequencies higher than the cut-off frequency. Both high-pass and low-pass filters can be used on video streams and digital image processing to perform transformations. This type of attack is viable on IPTV networks to remove some types of watermark.

## (viii) Transcoding

The process of translating an encoded video stream from one codec to a different codec may cause losses in the structure and quality of the video. Additionally, the digital marks may be lost during this process.

## (ix) Digital-to-Analog and Analog-to-Digital

A frequent mechanism for movie pirates is to bring a handheld camera into the cinema and record the movie. Although these are low-quality copies, they are still in demand in some markets. A similar situation could be presented with attackers filming the TV screen and then converting the recording into a digital video. This process may remove all digital marks from the video stream. Attackers may subscribe to a premium service, buy a high-definition LDC or plasma screen and record the image using a high-definition handheld camera.

## (x) Collusion Attack

If attackers manage to obtain a number of copies with different watermarks, they will use the available information to find the watermarks by comparing the different copies available. This situation can be present if two or more set top boxes from the same IPTV service provider are compromised. Intruders will have a number of copies of the same asset and will be able to compare them to extract or find the watermark. Attackers may be able to extract the sections of the stream that are 'different' and, by merging the common areas of a number of streams, clean the content of digital marks. Even with only two versions, the fields can be compared and an average value agreed, creating a new version free of digital marks.

## (xi) Mosaic Attack

This attack is based on taking the carrier image and dividing it into small parts that are then reconstructed for the viewer as a mosaic. The viewer will not notice any difference, but the pieces will be small enough for the marks to be lost. This attack is impractical for video streams.

## 4.5.5 Forensic Use of Digital Fingerprints

IPTV provides subscribers with high-quality digital video streams that are ready to be redistributed or burnt into a DVD. The type of content offered as part of the movie subscriptions, sports events and video on demand is likely to be of interest to the black market. Content owners and IPTV service providers have a significant problem as they have no assurances that content will be safe from illegal reproduction and broadcast.

In some cases, content will be recovered as part of police operations, seizing bootleg DVDs, or after tracing illegal broadcasting services. However, there is no immediate information allowing law enforcement agencies to identify the source of the material. From the initial recovery of the digital asset, anyone in the chain can be responsible for losing or stealing that particular material: employees working at the studio or content owners, translators or subtitle editors, couriers transporting copies, content aggregators, IPTV service providers, subscribers, etc.

There is always the question as to who was responsible for the content loss, and, with hundreds of IPTV service providers and content aggregators and millions of subscribers, this question is even more difficult to answer.

The IPTV infrastructure relies on the IP set top box to transform the encoded encrypted video stream and subsequently the MPEG contents into NTSC/PAL that can be interpreted by the TV set. Personal video recorders and similar technologies may also store copies in clear text outside the security domain provided by the DRM.

For those situations, content owners are starting to implement the use of digital marks to be used in forensic investigations. Digital fingerprints can be used to mark the video stream with information about the specific subscriber represented by the private key of the set top box, perhaps with a time stamp added. Once the content is marked, even if it is encoded, filtered or modified, the fingerprint will remain with the video stream and can be easily retrieved by forensic investigation using the keys embedded by the content owners. Information provided to the subscribers about the existence of this technology can reduce the inclination of some individuals to share contents on the Internet or sell them.

Some current set top box designs include chips with substantial computational resources capable of embedding a digital fingerprint in real time. This facilitates the process of marking the content and ensures that it reaches the set top box and is not intercepted en route.

Technology has been used by both content owners and criminals. Clearly, steganography is one of the next fields of combat, and both sides will try to prevail. Weak steganographic controls will end up in costly losses both in terms of credibility and in terms of digital assets. There are examples of steganographic tools that leave default headers and markings in digital files, making it trivial to recognize the existence of a hidden message on a particular file. Programmers and developers must ensure not only that the algorithms used are sound but also that the overall application used to embed the digital marks is secure and modifications are not easily traceable.

There are significant computing demands for set top boxes to determine the best locations to embed the digital mark. One viable option is to analyze the video stream at the head end while it is encoded and encrypted and add the ideal watermark locations to the metadata for the video stream. This metadata would be encrypted by the DRM along with the content and only when it reaches the set top box would it be used to embed the information about the serial for the set top box, subscriber info and any additional data.

## MPEG-2 Example

The following paragraphs are excerpts from a paper published by Anindya Sarkar, Upamanyu Madhow, Shivkumar Chandrasekaran and Bangalore S. Manjunath, Department of Electrical and Computer Engineering, University of California, Santa Barbara, CA 93106, USA (http://vision.ece.ucsb.edu/publications/sarkar_SPIE07.pdf) [14].

> The input video is decompressed at first into a sequence of frames, as shown in Figure 4.18 which provides a detailed block diagram representation of the video data hiding method. We embed data in the luminance ($Y$) component only. Here, we have employed the 'Selective Embedding in Coefficients' (SEC) scheme for hiding. In the SEC scheme, $8 \times 8$ DCT of nonoverlapping blocks are taken and the coefficients are divided by the JPEG quantization matrix at design quality factor. A uniform quantizer of step size $\Lambda$ is used on the DCT domain coefficients of the host image. Data is embedded through the choice of the scalar quantizer. The quantization index modulation (QIM) scheme uses even and odd multiples of $\Lambda$ to store 0 and 1 respectively.

Only those quantized DCT coefficients which lie in a certain low frequency band and whose magnitude exceeds a certain threshold are used for hiding – hence selective 'embedding' occurs. For a coefficient below the threshold, an erasure occurs. The embedded bit stream is then encoded and transmitted.

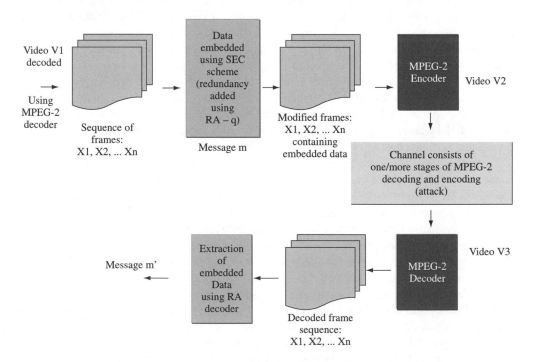

**Figure 4.18**   Video data hiding framework

The decoder does not know explicitly the exact locations where the data is hidden, but it uses the same local criteria (low frequency band for embedding and use of a magnitude threshold) as the encoder to guess the locations of hidden data. Channel noise may cause an insertion (decoder guessing incorrectly that there is hidden data) or a deletion (decoder guessing incorrectly that there is no data) which may lead to de-synchronization and decoding failure. We use turbo-like codes with strong error-correction capability and channel erasures at the encoder to account for this problem. Using the entire set of coefficients that lie in a designated low-frequency band, long code words can be constructed to achieve very good correction ability. We use repeat-accumulate (RA) codes in our experiments because of their simplicity and near-capacity performance for the erasure channels. A rate $1/q$ RA encoder involves $q$-fold repetition, pseudorandom interleaving and accumulation of the resultant bit stream.

Thus, once the data is embedded using the SEC scheme, error resilience is added using the RA codes. The modified frames are then converted to an MPEG-2 video. The video is subjected to MPEG-2 compression attacks (change in bit rate, variation in the Group of Pictures (GOP) size). Then, the received video is decoded to a sequence of frames, from which decoding (of the embedded data per frame) is performed iteratively using the sum-product algorithm.

In this paper, we have presented an adaptive hiding scheme where the embedding rate is varied for MPEG-2 video, depending on the type of frame (I/P/B) and on the MPEG-2 reference quantization parameter, which depends on the spatial activity in the macroblocks of a frame. As there are a variety of parameters at which the encoder may hide the data, the different parameter sets to be used need to be pre-decided and made known to the decoder. However, at runtime, the parameter set per frame need not be explicitly sent to the decoder since the RA decoding generally converges only if the decoder uses the right combination from its known set of parameters. Though this adaptive scheme is able to embed more data than the fixed-parameter based method and the frame error rate is also minimized, the temporal ficker and spatial noise terms are increased for the adaptive case. In our future work, we wish to incorporate methodologies to minimize the temporal ficker by including the per-frame distortion information in our frame-by-frame based hiding framework.

## 4.6 WWW? (What Went Wrong?)

### 4.6.1 Introduction

In recent years, several DRM and content protection mechanisms have been broken. This situation tends to be endemic and, to date, most widely accepted content protection mechanisms have been broken. One concept that tries to explain the problem basically refers to the fact that subscribers have the decrypting algorithm, the decrypting key and the encrypted content and we still expect this to be secure. This section will explore some examples of data protection mechanisms that have been broken. These examples are presented for historical purposes, as most of the problems have been fixed and new versions are available. In other cases, new media are expected to replace the vulnerable technology.

## 4.6.2  Satellite Television

Some information from the DirecTV site showing the current state of affairs (http://www.hackhu.com/) [15]:

DIRECTV's Office of Signal Integrity Presents:
Anti-Fraud and Anti-Piracy Enforcement Actions
THE TRUTH
This website provides the FACTS about DIRECTV's Anti-Fraud and Anti-Piracy
enforcement actions.
Lawsuits filed against over 25,000 fraud and piracy defendants.

News Update!
05/04/2007 Former North Carolina sheriff's deputy pleads guilty to racketeering and DIRECTV piracy. He faces 20 years in jail and $250,000 fine.

04/13/2007 End User Bruce Figler, represented by John Gibson, is found guilty of satellite piracy and ordered to pay DIRECTV $70,450.

04/05/2007 DIRECTV obtains judgment for $50,000 against Bounce (separate from Bounce Deuce), a NY bar, for Commercial Misuse.

03/20/2007 Jose N. Velasquez, a former installer, was arrested by the Carthage, Missouri Police Department for identity theft. He is being held on a $5,000 bond.

02/22/2007 Michael Hill pleads guilty to piracy charges and faces up to five years imprisonment and $250,000 in fines.

01/29/2007 DIRECTV obtains judgment for $50,000 against Bounce Deuce, a NY bar, for Commercial Misuse. See related posting on 12/27/06.

What is considered fraud?
The overwhelming majority of DIRECTV's customers provide us with truthful and complete information to establish and maintain DIRECTV service. Unfortunately, we have discovered instances in which individuals have used false information about their identity, fraudulent payment methods, the location of their receiving equipment, or the need for new access cards. We investigate such abuses and take appropriate actions, including pursuing DIRECTV's claims against individuals who access our service without proper authorization by or payment to DIRECTV. These claims may include civil litigation and/or forwarding investigations to law enforcement.

What are access cards?
The access card is a plastic card approximately the size of a credit card that is distributed to customers, either together with the receiving equipment or separately. Software contained in the access cards controls the receipt of DIRECTV programming. When a customer buys a satellite system with an access card from an authorized retailer, the customer is committing to subscribe to DIRECTV programming.

### Introduction

The security issues with satellite operators started a few years ago, similar to the decoders for cable TV services (unscramblers). Some new technologies started to appear, allowing unauthorized access to all the channels broadcast by satellite TV.

Original video streams broadcast by satellite operators were unencrypted. They relied on the smart card at the satellite receiver to authorize which channels to show and which ones to block. The smart card was an EEPROM card (Electrically Erasable Programmable Read-Only Memory). These EEPROM cards were easily modified, thereby lifting the restrictions over which channels to watch. Many of these cards were sold to people, allowing massive access fraud.

The solution came after more than a year, with the deployment of new cards with a dedicated ASIC (Application Specific Integrated Circuit) chipset. The ASIC module participated in the decryption of contents, allowing the satellite operators to broadcast encrypted streams. Once all valid subscribers had received the new cards, the satellite vendors moved to encrypted streams, killing all unauthorized EEPROM-based access. The ASIC card also included some basic functions (read, write, compare, etc.) that later became essential to deal with new attacks.

Being a more complex card, it had some security holes allowing access by intruders and facilitating modifications to the card. Intruders started to modify cards, and the satellite operators created new barriers. The process became a vicious circle that lasted for months. The satellite operators started to send updates to the receiver and downloaded a list of smart card IDs known to be used for access fraud. This killed many cards and forced attackers to write fake numbers on the EEPROM section of the cards. The numbers came from valid cards, and suddenly all cards used for access fraud were clones of valid ones.

After a few months, the satellite operators started to send a large number of updates that used the characteristics of the valid cards, checked the EEPROM areas of the card and created a loop on the cloned cards.

This modification affected only cloned cards. Valid subscribers had no alteration to the service. This measure was effective because the area written with the loop was part of the boot process of the card. At booting time, the card was no longer in a valid state and the whole system was not operational.

This process took many years, and a significant amount of money was invested in developing the technology and updates to it, in logistics and, recently, in court cases. As the site shows, 25 000 piracy and fraud defendants; clearly the costs are significant, and those are 25 000 people unlikely to become customers in the future.

Satellite receivers were not connected to the satellite operators. There was no option to authenticate and validate the set top boxes (satellite receivers). In IPTV environments there is a two-way communication, and there is always available information as to whom is receiving the contents. Broadcasts and access to VLANs by unauthorized users can be easily blocked.

## 4.6.3 DVD Protection

In 1995 the industry agreement on the DVD standard was reached, and from the original concept of information storage it evolved into a vehicle for content distribution. Content owners needed a mechanism to protect their digital assets being sold to consumers, and more specifically they needed a way to enforce access control and digital rights management. This technology was intended for both personal computers and stand-alone DVD players. Hence they required a simple technology to be deployed. The solution for this requirement was the creation of the DVD Copy Control Association (DVD CCA) responsible for licensing the content scramble system (CSS) to manufacturers of DVD hardware. Any DVD hardware maker would require a license from the DVD CAA to use CSS and be able to access DVD contents. This license would include the adherence to DRM requirements set by content owners and written on the disc, including region protection.

In general, commercial DVD discs have two main protections: region restriction and content encryption. Region restrictions are applied so a disc can only be watched by consumers in a particular region. The DVD player will check the region restrictions, and, if the disc and player have the same region flag, the content will be presented. DVD players on laptops have some flexibility in this matter, allowing consumers to change regions a handful of times before locking the region. This allows for cross-regional sales and travel. Some external DVD players for TV sets can be reprogrammed by consumers, which allows for some travel and cross-regional sales. CSS also blocks the skip and fastforward commands, which ensures that adverts and copyright notices cannot be skipped.

Content is also encrypted on the disc. This reduces the chances of unauthorized copies and modification. However, being a digital medium, bite-for-bite copies can be made, creating a true copy of the original and keeping the encryption protection intact.

### DVD CSS Exploit

In 1998 a group of open-source enthusiasts started to work on a software that could allow them to reproduce/play back DVDs on open-source operating systems. At that time, as claimed by the group, it was not possible to do that.

Through a collaborative effort, an initial piece of software was created that was capable of extracting a key and playing DVDs. However, that initial application had some problems with certain types of disk, and later, in 1999, a new application was released by a group of European hackers. This group was called MoRE (Masters of Reverse Engineering) and had a few members including two Germans and one Norwegian teenager (15 years old at the time). This software (DeCSS) was capable of unscrambling any DVD using a valid key taken from a DVD player for Windows. The operation of the software was very easy as it used a valid key to extract encrypted content.

From a security point of view, CSS failed in part owing to the architecture and design of the solution and also by the use of weak encryption (40-bit keys). The weak implementation of CSS by a DVD reader was responsible for losing one of the keys.

Once released, the DeCSS caused a major legal battle. Movie studios, via the Motion Picture Association of America (MPAA) and the DVD CAA, started a legal battle to stop the CSS information from being disclosed. They claimed that the CSS functioning was a trade secret, and the situation reached 'movie-like' levels when people started to wear T-shirts with the DeCSS code and send emails with the DeCSS as a signature; even a website was created to promote 42 ways to release the code. In practical terms, once the CSS was broken there was little hope of maintaining the integrity of DVD protection. Over time, the Internet became full of software allowing unrestricted access to DVDs, and, with the vast amount of titles out there, and complexities of backward compatibility, it is difficult to see a CSS 2.0 being released.

## 4.6.4 AACS on Blue-Ray and HD-DVD

A new generation of disc protection mechanisms was the advanced access control system (AACS) administered by the AACS Licensing Administrator who provides the licenses to Blue-Ray and HD-DVD manufacturers.

AACS works by encrypting the contents using AES 128-bit keys, a major improvement over the proprietary 40-bit cipher used by CSS. Brute-forcing AACS is out of the question and leaves only the option of attacking the implementation of the algorithms.

AACS provides individual disc players with a unique set of decryption keys. If a successful attack is disclosed, then the keys to a particular player can be revoked. Future content will be issued with specific rules that block access by the revoked reader.

Even if the successful attack is not disclosed, movies protected by AACS are divided into different sections that have been encrypted with different keys. Analyzing the digital watermarks on the different keys enables the AACS LA to ascertain which player has been compromised and revoke access to that type of player.

Much like the digital video wars at the end of the 1990s, AACS provides a basic mechanism to block certain types of access; this corners attackers and facilitates future control.

AACS uses a number of elements to deploy the security mechanisms. Some basic elements are as follows.

### (i) Volume IDs

Unique identifiers stored on discs. VIDs cannot be duplicated using standard recordable drives, and hence bit-by-bit copies are not possible. This increases the complexity of bit-by-bit piracy.

VIDs are required to read discs, but there are indications of a technique to avoid this security mechanism.

### (ii) Content Decryption

AACS embeds four elements on the disk: the master key block (MKB), volume ID, encrypted key and encrypted content. MKB is encrypted and each drive will use its private keys to recover the MKB.

The MKB provides the master processing key (Km). By combining the Km and the volume ID (using a one-way function), the software obtains the volume unique key (Kvu). The Kvu is used to recover the encrypted title key, which in turn is used to recover the content. The process of content decryption is illustrated in Figure 4.19.

### (iii) Audio Watermarking

AACS can support the use of audio watermarking. The audio track of the content would be marked. AACS complaint readers are required to check that the watermark is present during playback. Failure to find the watermark on the audio channel will cause the reader to lock the signal and playback will be stopped. This approach will protect against manual recordings made using high-definition camcorders and similar media. The sound track from that recording will not include the watermark.

### (iv) Fair Use

AACS includes a functionality that supports fair use. Consumers are allowed to make controlled copies of the contents. These copies will retain the DRM controls, but they will be available on different equipment used by the consumer. One example would be to store a copy of a disc on the home entertainment system or use wireless to broadcast content to a handheld device in the home.

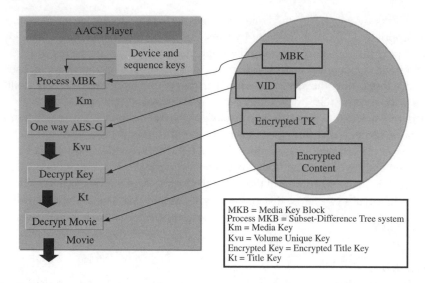

**Figure 4.19**   AACS process

*(v) Image Constraint Token*

A flag included in the DRM protection of the content may restrict the quality of playback on analog media. This is known as the image constraint token (ICT). Once enabled, the disc player will allow analog outputs only on $960 \times 540$ bits. For high definition ($1920 \times 1080$ bits), the output requires to be compatible with high-definition copy protection.

*(vi) Exploit*

Both title keys and one of the keys used to decrypt them have been extracted from memory positions used by PC software accessing AACS compatible drives, and additionally some device keys (used to calculate the processing key) and a host private key (a key signed by the AACS LA, used for handshaking between host and HD drive; required for reading the volume ID). There are a number of websites on the Internet currently hosting title keys. One site has a collection of more than 85 different title keys.

The main vulnerability exploited to date is the fact that software players need to have the keys in memory available to decrypt the content. There are no secure hardware repositories within the PCs, no place where keys can be stored temporarily while they are used. This allows attackers to capture keys while in use.

## 4.6.5 Videos Over the Web

A British broadcaster decided to stop offering web downloads of videos owing to the news regarding known vulnerabilities of the DRM application used by them. This was an incipient service and did not have a significant impact on subscribers. However, it showed how a successful attack on a DRM platform may shut down a video service. A similar situation could occur on IPTV if appropriate measures are not taken while designing the security.

This particular DRM solution used a 56-bit symmetric key to encrypt contents, and, as part of the header, it had a key identifier and a URL for license requests. When subscribers wanted to access the file, the system verified in the internal key database for the specific key identifier; if the key was not available, it proceeded to check the URL for a valid license.

Licenses had an XML format signed by the content owner or content provider. This XML file had the DRM restrictions: play once, expiry date, etc. Licenses also included the encrypted key required to access the content.

The attack used for this particular DRM platform was based on accessing the memory of the computer and extracting the 56-bit content key. This is a similar approach taken to break HD-DVD and Blue Disc protections.

## 4.7 Authentication

Access to contents and services must be provided only to authorized users. For this there are specific authentication mechanisms that can be used. IPTV has significant advantages compared with satellite, disc or out-of-band transmissions. IPTV service providers can validate each one of the set top boxes and ascertain if those particular elements represent authorized subscribers or not. IPTV does not have an open environment where all interested parties can request content. DRM is not the only line of defense.

IPTV environments have network authentication and authorization as the first line of defense. This is done at the DSLAM level, and the second line is provided by the middleware server and authentication, relying on DRM to provide the third level of protection.

*(i) Network Authentication*
The initial authentication takes place at the network level using a broadband remote aggregation server (B-RAS). For small networks, a centralized B-RAS can be used. It works by controlling all the traffic, but it becomes a bottleneck for larger networks.

One of the options for larger networks is to use a distributed B-RAS, with more scalability and reliability than a centralized architecture.

Another option is the use of an integrated B-RAS. This merges some functionalities of the B-RAS into the digital subscriber line access multiplexer (DSLAM), which reduces the implementation costs and reduces the traffic generated.

The dynamic host configuration protocol (DHCP) can also be used to perform the authentication of new elements, which eliminates the requirement of a B-RAS.

*(ii) Middleware Authentication*
Set top boxes must provide credentials and authentication data to the middleware server. These can be in the form of a challenge response pair, digital certificate and digital signatures or one-time passwords. The middleware will verify internally if the set top box represents a paid subscriber and will also provide information to the subscriber (and the internal IPTV systems) about the packages that have been authorized for a particular subscriber.

## 4.8 Summary

Intellectual property rights and management are a relatively new concept within legal frameworks. Kings and nobles supported artists and authors, giving them grants and financial support. However, there was no legal support for licensing or expectation of any revenues from the reproduction of their work.

The civilized world created legal frameworks to ensure that authors of creative works could have the rightful expectation to benefit from their intellectual work. Patents, copyright, trademarks and licensing mechanisms were established, allowing sustainable business models.

Within IPTV environments, intellectual property rights are enforced using digital rights management technologies. These rely on encryption algorithms to ensure that subscribers are required to follow the licences provided by the content owner via the IPTV service provider.

Symmetric and asymmetric encryption are used within public key environments digitally to sign contents, authenticate components and exchange encryption keys. PKI can be used to authenticate set top boxes each time they join the network or when they attempt to access the middleware server or any other component on the head end. Additionally, encryption keys for video on demand can be sent to the set top box using the public key from the set top box, thus ensuring that only valid subscribers have access to the content.

IPTV service providers can deploy steganography to mark video feeds, embedding information about the set top box used to view specific contents. This information can be used at a later stage during forensic examinations of recovered content. Establishing the source of the content would help investigators to close open systems. This technology can also be used as a deterrent, reducing the number of subscribers willing to record content.

DRM technology is not perfect, and, although the algorithms used are very strong and reliable, implementations have been broken in the past. Some examples of broken DRM systems include the one used by satellite TV, DVDs and music files. When choosing a DRM system, security professionals must be satisfied that the vendor has followed secure processes for programming the application and it has been tested by independent reviewers.

# References

[1] 'The United States Constitution', 1788. Available online: http://www.house.gov/house/Constitution/Constitution.html [2 October 2007].
[2] US Copyright Office, 'The Digital Millennium Copyright Act', 1998. Available online: http://www.copyright.gov/legislation/dmca.pdf [2 October 2007].
[3] US Supreme Court, 'White-Smith Music Publ. Co. v. Apollo Co., 209 US 1', 1908. Available online: http://caselaw.lp.findlaw.com/cgi-bin/getcase.pl?court=US&vol=209&invol=1 [2 October 2007].
[4] US Congress, 'The International Copyright Act – Amendments to the 1870 Copyright Act'. Available online: http://www.ellengwhite.info/copyright_law_us_1891.htm [2 October 2007].
[5] Federal Information Processing Standards Publication 46-2, Data Encryption Standard, 1988. Available online: http://www.itl.nist.gov/fipspubs/fip46-2.htm [2 October 2007].
[6] RSA, 'RSA Algorithm'. Available online: http://www.rsa.com/rsalabs/node.asp?id=2146 [2 October 2007].
[7] Federal Information Processing Standards Publication 197, Advanced Encryption Standard, 2001. Available online: http://csrc.nist.gov/publications/fips/fips197/fips-197.pdf [2 October 2007].
[8] International Telecommunication Union, ITU-T Recommendation X.509, 1997. Available online: http://www.itu.int/rec/T-REC-X.509-199708-S/e [2 October 2007].
[9] Netscape Corporation, 'Introduction to SSL', 1998. Available online: http://docs.sun.com/source/816-6156-10/contents.htm [2 October 2007].
[10] Internet Engineering Task Force, 'TLS Charter', 1996. Available online: http://www.ietf.org/html.charters/tls-charter.html [2 October 2007].
[11] Villegas A., 'DRM Convergence Analysis of Products and Standards', Alcatel-Lucent, 2007.
[12] Meerwald, P., 'Digital Image Watermarking in the Wavelet Transform Domain', Master's thesis, Department of Scientific Computing, University of Salzburg, Austria. Available online: http://www.cosy.sbg.ac. at/~pmeerw/Watermarking/MasterThesis/ [2 October 2007].

[13] Komatsu, N. and Tominaga, H., 'A Proposal on Digital Watermark in Document Image Communication and its Application to Realizing a Signature', *Electronics and Communication in Japan, Part 1 (Communications)*, **73**(5), 1990, 22–23.

[14] Sarkar, A., Madhow, U., Chandrasekaran, S. and Manjunath, B.S., '*Adaptive MPEG-2 Video Data Hiding Scheme*', Department of Electrical and Computer Engineering, University of California, Santa Barbara, CA 93106, USA. Available online: http://vision.ece.ucsb.edu/publications/sarkar_SPIE07.pdf [2 October 2007].

[15] DIRECTV's Anti-Fraud and Anti-Piracy Enforcement, 2006. Available online: http://www.hackhu.com/ [2 October 2007].

## Bibliography

Office of Technology Transfer, University of Oregon, 1999. Available online: http://www.uoregon.edu/~copyrght/Docs_Html/XII_XIII.htm [2 October 2007].

DeLong, B., 'Economics of Intellectual Property', 2007. Available online: http://delong.typepad.com/sdj/economics_intellectual_property/index.html [2 October 2007].

Varian, H.R., 'File-Sharing is the Latest Battleground in the Clash of Technology and Copyright', *The New York Times*, 2005. Available online: http://www.ischool.berkeley.edu/~hal/people/hal/NYTimes/2005-04-07.html [2 October 2007].

Lessig, L., 'The Creative Commons', *Florida Law Review*, 2003. Available online: http://homepages.law.asu.edu/~dkarjala/OpposingCopyrightExtension/commentary/LessigCreativeCommonsFlaLRev2003.htm [2 October 2007].

# 5

# Existing Threats to IPTV Implementations*

Co-authored by Andrew R. McGee, Frank A. Bastry and David Ramirez

The case study of the security of typical IPTV networks revealed a number of critical threats that could cause security incidents in the future. The number of potential vulnerabilities found was also very high, which shows the importance of analyzing both the design and subsequent implementation of an IPTV service.

An evaluation of the IPTV at the home end and of the DSL access network identified 69 relevant information assets. These assets were distributed across the IPTV application, services and infrastructure layers as illustrated in Figure 5.1. The number of information assets contained in a specific IPTV service offering may vary because different IPTV service providers will have unique network architectures determined by a number of factors such as type of access network (e.g. DSL, cable, fiber), preferred implementation model (e.g. nPVR or CPE-based PVR), etc. From the design and deployment point of view, security professionals must have complete visibility of each and every component that will become part of the IPTV service offering, as well as any potential vulnerabilities in the service offering.

Industry fora and standards organizations were consulted for known threats to IPTV services, and the priorities of these threats were assigned on the basis of the relationship of each threat to the IPTV service provider's revenue and reputation. For example, network outages directly impact upon the service provider's revenue, and theft of content directly impacts upon the service provider's reputation with content providers. A vulnerability analysis was performed to determine how vulnerable each of the information assets is to the threats. The risk of the vulnerability is based on the priority of the associated threat. The vulnerability analysis showed approximately 478 potential vulnerabilities, with

---

* This chapter is based on the research work by Bell Laboratories on IPTV security. As part of the research, a case study was performed that identified the threats to and vulnerabilities of typical IPTV networks.

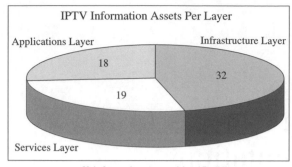

69 information Assets identified

**Figure 5.1**   Home-end identified assets

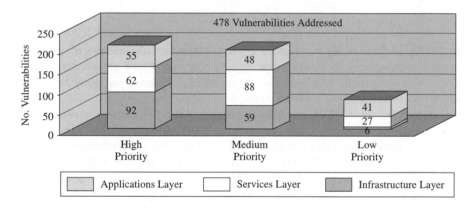

**Figure 5.2**   Vulnerabilities found

approximately 209 high-risk vulnerabilities. As illustrated in Figure 5.2, these vulnerabilities were distributed towards the medium and high category.

Vulnerabilities ranged from potential theft of credentials within the set top boxes to denial of service attacks caused by errors in the configuration of the DSLAM.

A similar approach was undertaken for the IPTV VOD and DRM systems, which included content management, content distribution and conditional access functionality. This portion of the IPTV network was found to contain 101 information assets. As illustrated by Figure 5.3, the assets were evenly distributed across all three layers.

The home end shares a number of vulnerabilities with the VOD and DRM systems, mostly related to the protocols used to transport information as well as some of the files that have to be exchanged between parties. As Figure 5.4 illustrates, the VOD and DRM systems have a significant percentage of high-risk vulnerabilities. This is primarily due to the functionality of the system (e.g. interface to content providers) and the number of subscribers that would be affected by a successful attack against one of these systems.

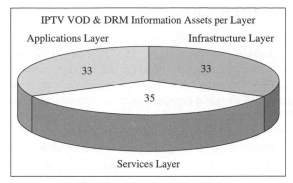

Figure 5.3   VOD and DRM identified assets

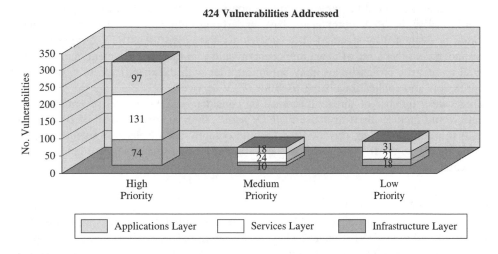

**Figure 5.4**   Vulnerabilities found

These results clearly show that there are a high number of potential security problems that may affect any IPTV service. Individual IPTV service providers will deploy network architectures suited to their own needs and the video services they offer to their customers. Therefore, the number of vulnerabilities may change owing to the uniqueness of the IPTV network being deployed. Each IPTV network must be evaluated in a similar manner, analyzing the design of the solution and the security characteristics of all its components in order to identify potential vulnerabilities and the actions that must be taken to control the risk. Following the approach of a layered model, all layers must be reviewed to establish the security posture of the network before the IPTV service becomes operational.

All complex systems have similar situations. Hundreds of vulnerabilities could be found during a security analysis of a VoIP solution or the high-speed Internet service; what is important is to know the vulnerabilities and match them with controls.

## 5.1 Introduction to IPTV Threats

IPTV inherits all the security vulnerabilities from the network used for transport; as such, it would keep the vulnerabilities present on TCP/IP and within the transport network components. Therefore, the same basic measures required to protect any other TCP/IP network apply to the IPTV infrastructure. This is particularly true as many IPTV providers will use existing Internet links to send contents, or their infrastructure will be used for several services including IPTV, VoIP, hosting and Internet access services.

There are some critical characteristics of IPTV-related data that should be protected. Video and sound are sensitive to delays and packet loss, and, even if the protocols and applications have some built-in resilience, bad communication will certainly affect the customer experience.

Special mechanisms should be implemented to control the TCP/IP-related vulnerabilities and ensure the proper operation of the IPTV service. IPTV components have specific characteristics that should be considered during the security evaluation. For example, middleware services tend to run over HTTP/HTTPS, and there are many known vulnerabilities that could be exploited to take control over the middleware server.

Additionally, the IPTV components have underlying vulnerabilities that are not tracked, reported or evaluated by standard IT security tools and procedures. IPTV service providers must understand that a set of new components would join their environment, and operators, as well as security professionals, may not be familiar with the specific security issues with the protocols and applications being used for IPTV services.

There are a number of threats that must be considered. A high-level inventory of such threats would include: theft or abuse of IPTV assets, theft of service, theft of IPTV-related data, disruption of service, privacy breaches and compromise of platform integrity. Known security threats to the IPTV service can be classified into those categories. Figure 5.5 has a representation of the main threats to IPTV environments.

*(i) Theft or Abuse of IPTV Assets*
The main business assets of the IPTV service are the digital copies of content stored and transported within the infrastructure. These assets are at risk of being stolen by third parties as well as being modified, affecting the ability of the IPTV service provider to derive value from the asset. Abuse of IPTV assets is related to manipulation of the IPTV infrastructure for purposes outside the planned functions of the elements.

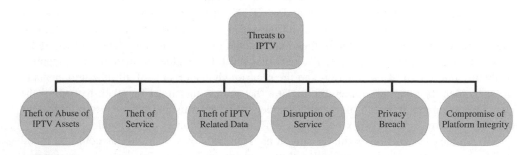

**Figure 5.5**  IPTV threat hierarchy

Some examples of this threat would be unauthorized access to the video repository to steal digital copies of assets, unauthorized copies of digital media storing movies or unauthorized access to the transcoder equipment to steal digital copies.

Theft of IPTV assets is done with the intention of resale, distribution and modification and, in general, is part of a larger scheme to benefit financially from the assets. If a subscriber intends to receive contents without paying, or an external party wants to receive an IPTV broadcast without paying, this would be considered theft of service.

*(ii) Theft of Service*

Any activity where the end-user is receiving IPTV services without the proper level of subscription will be considered theft of service. This will include system errors or minor modifications that increase the package by a few channels as well as VOD access without proper subscriptions.

In some situations there will be an individual responsible for theft of assets (stealing a digital copy of a movie and broadcasting it) and many other individuals responsible for theft of service. Both will have different countermeasures.

This type of threat comes from manipulation of the middleware and provisioning applications, allowing subscribers to add channels to their subscription without actually paying for it. In some cases this situation can be presented if the VLANs and access controls are not properly deployed. Intruders may be able to redirect a video broadcast stream to unauthorized set top boxes that use a valid license from a different TV package. Modification of the software inside a set top box to reprogram it to access blocked channels is another possible way to steal service.

*(iii) Theft of IPTV-related Data*

The IPTV environment includes a significant amount of data related to subscribers, digital assets, infrastructure and the service. Subscriber data is considered to be within the privacy threats and is also covered by legislation in most countries. IPTV-related data may be used by competitors and criminals or for eavesdropping. This information must be protected to avoid brand damage and loss of customers.

IPTV-related data is stored in most components of the head end and aggregation network. For example, the middleware would contain information about the subscriber preferences, and the DSLAM may include details of the video requests. Account information and billing information would be included on the business-related servers and could be used for identity theft and related crimes.

*(iv) Disruption of Service*

Subscribers are used to having an 'always on' service with consistent quality levels. Satellite, terrestrial and cable TV services tend to be reliable and free from interference; customers expect high levels of quality from TV and will not accept frequent interruptions or bad-quality images, especially when paying for it. There is a clear threat of disruption of services from attacks to the head end affecting all subscribers; there are also threats of regional or localized attacks that would affect a few thousand subscribers. Reliability of the service should be added to the network architecture requirements in order to ensure a consistent service to the majority of subscribers.

Under certain circumstances, intruders may be able to take control of a significant number of set top boxes. Intruders may be able to direct a denial of service attack targeting the middleware server and blocking access to the IPTV service. Other attacks may include changing the configuration of the DSLAM, switches or multicast rules to cause a disruption on the service.

*(v) Privacy Breach*
Privacy laws and regulations establish the responsibility of IPTV service providers to protect personal information from subscribers. Intruders may be able to have access to database servers storing personal information or capture transactions from set top boxes.

Mechanisms should be deployed to avoid disclosing personal information from subscribers, including the encryption or obfuscation of records.

*(vi) Compromise of Platform Integrity*
The integrity of the platform delivering the IPTV service must be ensured to avoid escalating security incidents. If an intruder is able to compromise the integrity of the platform, he or she may be able to escalate this attack and take over larger areas of the service.

Intruders can use the web service from the middleware server to take control over this component and escalate the attack by connecting to other components on the head-end network. Appropriate controls should be deployed to ensure consistent platform integrity and timely detection of intrusions.

## 5.1.1 Specific Threats to IPTV Environments

Security for the IPTV service will rely on the security of the underlying infrastructure. Users and content providers are now exposed to the same security problems faced by any computer user and any Internet service.

If a computer worm infects a home PC, then this self-replicating program will try to infect the STB and any other equipment connected, including the head-end equipment. In a sense there is a logical path between the Internet and the head end. In the past there was absolutely no way a hacker could disrupt standard TV services. With IPTV that situation is a distant possibility, but a possibility nevertheless.

Previous industry practices with the telecom industry show that home-end equipment tends to be configured with minimum security levels including using a standard password for all equipment or leaving the default values set by the hardware vendor. While this facilitates installation and maintenance, it also creates several security problems.

Some of the general infrastructure security threats are as follows:

- *Unauthorized access to elements.* Intruders exploiting known vulnerabilities or default passwords to have unauthorized access to a particular system.
- *Denial of service (DOS).* Intruders could exploit known security problems with the applications or stack implementations to crash a particular system. Another known approach is bandwidth consumption or resource exhaustion. Intruders will send a very large number of requests, causing either the networking equipment or the application to stop working owing to excessive load.
- *Operating system vulnerabilities.* All operating systems have security vulnerabilities that can be used one way or the other to affect the operation of the service.
- *Application attacks.* The IPTV market has several applications currently used by content providers. Intruders might discover security problems with the applications and cause problems with the availability of the service.

### Theft or Abuse of IPTV Assets

Theft of content (some apply also to theft of service):

- capturing the digital certificate from an STB to order contents and even broadcast/ redistribute the stream to other subscribers;
- packet capture on the home network and IP subnet;
- output from an analog output port to an external recording device;
- output from a digital port to an external recording device;
- implementing the playing of more than the number of allowed plays;
- accessing illegitimate content (e.g. pirated content);
- circumvention of conditional access systems (CASs) to enable access to content;
- copying content from the disk storage on a video server or STB.

### Theft of Service

Theft of service (some applies also to theft of IPTV assets):

- unlawful taking of a benefit from an IPTV service provider, intended to deprive the IPTV service provider of lawful revenue;
- defrauding an IPTV service provider;
- unauthorized deletion or alteration of billing information;
- STB or smart card cloning;
- implementing VOD trick play and long pause functions for more than a specified period of time;
- circumvention of conditional access systems (CASs);
- massive replication/dissemination of information, enabling theft of service.

### Theft of IPTV-related Data

- unauthorized access;
- theft of subscriber records;
- theft of configuration data;
- theft of metadata.

### Disruption of Service

*(i) Traffic/Packet Flooding*

- DOS attack on a user endpoint by sending a large number of valid packets, causing interruption of service, some of which may affect network elements as well; application stops owing to overload.
- endpoint packet flooding scenarios cause network element, video server or gaming server to crash, reboot or exhaust all resources;
- DOS – bandwidth consumption or resource consumption; high volume of traffic (e.g. to a multicast group);
- potentially impacting upon thousands of subscribers (e.g. DSLAMs and video servers support thousands of subscribers).

*(ii) Malformed Packets and Messages*
- disabling endpoints with invalid messages – DOS attack on the endpoint (e.g. STB, video server) by sending a number of invalid messages that could cause the endpoint to crash, reboot or exhaust all resources;
- malformed protocol messages – sending of malformed protocol messages (e.g. messages with overflow or underflow) to the device which degrades its performance to the point of being unable to process normal messages;
- malformed messages that cause buffer overflow;
- potentially impacting upon thousands of subscribers (e.g. DSLAMs and video servers support thousands of subscribers).

*(iii) Spoofed Messages*
- DOS attack that disrupts IPTV service by causing an IPTV session to end prematurely;
- spoofing of control messages; malicious control traffic injected into the communications, causing applications or video servers to malfunction or traffic to be sent to the wrong destination; forged control messages used to alter the structure of multicast distribution trees and affect the data distribution across them; DOS – bogus broadcast message claiming there is a high loss rate on the channel or high congestion (the source will reduce the transmission rate, affecting other subscribers);
- forged IPTV subscriber messages and application or video server responses;
- change IP and MAC addresses to spoof other user MAC and IP addresses to capture IPTV streams.

*(iv) Underlying Platform DOS*
- vulnerabilities of the underlying operating system or firmware on which the application or service runs;
- 'point-and-shoot' exploits freely available for download on the Internet;
- hacking on operating system and software supporting games, video storage functions, etc.
- DOS attacks reducing the performance of the device;
- exploitation of these vulnerabilities has the potential to propagate to thousands of devices (e.g. every STB);
- potentially resulting in the redeployment of or maintenance to thousands of devices (e.g. every STB).

**Privacy Breach**
- invasion of subscriber privacy and eavesdropping;
- call pattern tracking to discover identity, affiliation, presence and usage;
- traffic capture – unauthorized recording of traffic, including packet recording, packet logging and packet snooping (includes management and signaling traffic);
- unauthorized access to subscriber media stream;
- unauthorized access to management traffic;
- unauthorized access to signaling traffic;
- information harvesting – an unauthorized means of capturing identity that enables subsequent unauthorized communication and theft of information (consists of the collection of IDs, which may be numbers, strings, URLs, etc.);

- media reconstruction – unauthorized monitoring, recording, storage, reconstruction, recognition, interpretation, translation and/or feature extraction of any portion of a video communication including identity, presence or status;
- unauthorized disclosure of subscriber IPTV capabilities;
- unauthorized disclosure of a subscriber's previous or current IPTV usage or activities (e.g. subscriber viewing history of broadcast or VOD content, online gaming activities, etc.);
- replay attacks involving media (replaying captured media for malicious gains, or invading privacy by replaying media for personal use).

*(i) Interception and Modification*
- conversation impersonation and hijacking – the injection, deletion, addition, removal, substitution or replacement or other modification of any portion of a communication with information that alters any of its content and/or the identity, presence or status of any of its parties (includes management and signaling traffic);
- unauthorized access, modification or deletion of electronic program guide (EPG) information;
- hijack video stream – insertion, modification or deletion of a video stream in an unauthorized manner;
- SPIV (SPAM over IPTV) – displays unsolicited pop-up advertisements;
- unauthorized broadcasting of material (for political or other reasons).

*(ii) Compromise of Subscriber Application Data*
- unauthorized disclosure, creation, modification, duplication or deletion of data created and/or used by subscriber-accessible applications;
- includes information stored in the service provider's network on behalf of subscribers (e.g. video content recorded by nDVR).

*(iii) Compromise of Subscriber Information*
- social engineering to obtain subscriber information;
- unauthorized disclosure, creation, modification, duplication or deletion of subscriber information (e.g. address, phone number, account number, credit card information, DNS entries, etc.);
- limited to individual subscribers.

**Compromise of Platform Integrity**
*(i) Misrepresenting Authority and Rights*
- presentation of a false authority as if it were true with the intent to mislead;
- presentation of a password, key or certificate of another (e.g. video server or content management system administrator);
- unauthorized acquisition and use of subscriber service related authentication information (e.g. user ID/password, session keys) (limited to individual subscribers);
- unauthorized acquisition and use of administrative authentication information (e.g. user ID/password);
- replay attacks involving signaling.

*(ii) Compromise of Installed Software, Service-related Data or System Configuration*
- malware, spyware or rootkit insertion;
- unauthorized duplication, installation, alteration or deletion of production software and configuration files;
- unauthorized duplication, disclosure, creation, modification or deletion of service-related data (e.g. system logs, billing information, decryption keys, storage containers for decryption keys, etc.);
- D-DOS using compromised devices to crash the IPTV service;
- unauthorized creation or modification of subscriber service related information (e.g. authentication information, session keys);
- unauthorized or unnecessary activation/deactivation of logical (protocol) ports.


*(iii) Resource Exhaustion*
- deficiencies in software or hardware that cause depletion of memory resource (e.g. buffers) in a network element;
- deficiencies in software or hardware that consume most CPU resources in a network element;
- hardware or software errors that limit available bandwidth of a communication link;
- deficiencies in software or hardware that generate unnecessary messages reducing bandwidth resources;
- for example, infinite software loops or routing loops.


*(iv) Unauthorized Network Scans and Probes*
- port scanning/ping sweeps – attackers can run publicly available scanning software on a host with connectivity to an IPTV network, and host services on devices monitoring the ports will respond, potentially providing information to the attacker;
- vulnerability scanning (e.g. nessus) and network mapping (e.g. NMAP) – attackers can run publicly available software on a host with connectivity to an IPTV network that queries the device configuration and network topology;
- unauthorized remote access to software or functions resident on a device (e.g. utilizing a rootkit to provide a backdoor).


*(v) IPTV Session Hijacking and Service Masquerading*
- impersonation of a legitimate IPTV service provider – capturing a digital certificate from a provider to modify streams and include any information desired;
- impersonation of a legitimate network device, video server, gaming server or DRM server;
- man-in-the-middle attack;
- redirection of a video stream to an unauthorized device.


*(vi) Unauthorized Management*
- unauthorized use of an on-board management application or execution of management commands (e.g. manipulation of a modem configuration to block specific services);
- forged/modified management protocol messages (e.g. manipulation of a modem configuration to block or allow specific protocols, such as SNMP).
- modification of remote management messages (e.g. MITM);

- illegitimate subscriber self-provisioning actions (for example, reconfiguring an STB to remove bandwidth limitations in order to produce slow connections for other subscribers or increase bandwidth for yourself);
- authorized management agent performing unauthorized activities;
- unauthorized content management (e.g. loading, deleting video content or modifying the trigger date – the date that content becomes available to the viewing public);
- unauthorized subscriber management (e.g. unauthorized subscriber provisioning activities including upgrade/downgrade of subscriber viewing privileges).

The following sections include a more detailed view of the threats within the head end, the aggregation/transport network and the home end. This chapter could not include details on specific platforms, products or vendors as, with time, the information would lose its value as a viable alternative. The chapter includes threats, topics, questions and examples that security professionals working on IPTV should be aware of. This information can be used as a template and be applied against an existing or proposed IPTV environment to determine its security posture.

## 5.2 IPTV Service Provider – Head End

The IPTV environment will have different sets of threats depending on the function. The impact of security incidents on the head end is naturally greater than that of security incidents on the home end. The aggregation and transport network will have a number of threats on account of the shared services being supported. The following sections will present examples of vulnerabilities on the different sections of the infrastructure. The sections are represented in Figure 5.6.

Within the head end, it is important to consider the threats posed by IPTV service provider personnel. Systems resident in data centers or central offices are typically more exposed to internal than to external threats, and sometimes employees have access to more information than they really should. In the past there have been countless cases of employees defrauding companies, and this includes utility and telecommunication companies.

Internal applications must be set up in such a way that they restrict access to authorized users only and internal modifications are always logged and confirmed by a second person. This reduces the chances of internal fraud.

**Figure 5.6**  High-level IPTV environment

The same risk is posed by employees and social engineering attacks. A perfect example is the 'Paris Hilton sidekick hack', where a group of teenage hackers in California called a store and tricked one of the clerks into disclosing information about an internal site with phone numbers and private client information, as well as the password and user name to access the site.

The hackers then exploited a vulnerability of the mobile company's website, managed to reset the account and gained complete access to private pictures and contact details.

Companies invest millions of dollars implementing a secure infrastructure and monitoring hardware and software tools. However, call center operators are usually low paid and receive minimum training, creating the perfect opportunity for hackers to obtain private information.

Some specific threats within the head end are as follows.

## 5.2.1 Video Feeds – Live or Prerecorded (Physical Media, OTA, etc.)

There is a threat of disruption of service. This asset is exposed to destruction of physical media. Careless handling of physical media can result in its destruction. Service disruption of live video feed can be caused by transport network or equipment malfunction. Live video feed carrying content can be interrupted by a network malfunction or equipment malfunction in the carrier's network.

There are a number of threats affecting platform integrity. An IPTV service provider may use illegitimate content by mistake or after a deliberate modification by one of the operators. This scenario includes the alteration of legitimate prerecorded content.

The IPTV service provider may receive video feed from an illegitimate content provider, including a modified feed with inappropriate content.

Additionally, there could be physical damage to media, affecting sections of the content. Digital copies can be damaged in transit or while stored.

There are some risks of theft/removal. There could be unauthorized replication or removal of physical media. IPTV service operators with access to digital media could replicate contents. In most cases this would be an internal attack.

## 5.2.2 Video Switch

There is a threat of disruption on this asset. Unauthorized access to the video switch can result in accepting an unauthorized video feed for loading into the IPTV service. The unauthorized video switch operation would most likely be the result of an internal attack.

Unauthorized management commands may stop the video feed reaching the rest of the IPTV components and cause an interruption in the service to subscribers.

Overwriting of video switch software or data via a buffer overflow could cause the video switch to hang, crash or terminate prematurely.

The video switch has some additional threats related to theft of IPTV assets. Unauthorized access to the video switch can result in switching a legitimate video feed to an unauthorized recording device for the purpose of unauthorized use or distribution. This would most likely be an internal attack.

## 5.2.3 Ingest Gateway (Video Capture)

The ingest gateway is exposed to some threats related to theft/removal. Unauthorized access to the ingest gateway can result in storing ingested content on an unauthorized storage device (e.g. removable media) for the purpose of unauthorized use or distribution. This would most likely be an internal attack as most ingest gateways are too complex to operate via the network.

There are additional threats related to disruption of service. Unauthorized access to the administrative console can be used to send commands over the content management network that halt the ingest gateway (e.g. a kill command).

Overwriting of ingest gateway software or data via a buffer overflow could cause the ingest gateway to hang, crash or terminate prematurely.

## 5.2.4 Platform SW/OS – Stored/Running

This applies to all operating systems supporting IPTV-related applications.

All operating systems are exposed to threats related to disruption of service. Computer worms and viruses as well as directed attacks can be aimed at causing denial of service within the IPTV platform. Unauthorized access can result in the deletion of files required by the operating system. Malformed packets or unauthorized requests can be sent to operating systems and cause critical files to be removed. Operating systems have many services running on open ports. Without proper patching and configuration, intruders may be able to use those ports to shut down the service.

Computer worms and viruses could be used to infect a large number of servers within the head end and disrupt the service.

There are additional threats to the integrity of the platform. Under certain circumstances, there could be an unauthorized modification of platform software/OS configuration files. Intruders may be able to plant a rootkit and modify files or parameters within the operating system.

Operating systems control access to the files within the local drives. Unauthorized browsing can occur after a system break-in by an intruder (external attack) or by an insider gaining unauthorized access to the log files or information stored on the system.

## 5.2.5 Content Management System

The content management system is exposed to a number of threats including platform integrity threats related to buffer overflow vulnerabilities. This type of vulnerability arises owing to poor security checks on the original application code. Intruders can exploit these vulnerabilities by input strings longer than the allocated buffer. Modifications to the content management system may allow intruders to plant backdoors or malware on the system and obtain copies of digital assets.

Intruders can cause disruptions of the service. They can send protocol messages that would cause the application to halt, or may be able to send management commands shutting down the application. Management commands may also be executed by an intruder able to perform management activities. Overwriting of content management software or data via a buffer overflow could cause the content management software to hang, crash or terminate prematurely.

There are some threats related to theft of IPTV assets. The content management software may be manipulated by intruders to obtain copies of digital assets. This is one of the few systems allowed to communicate with the video repository and as such is a perfect stepping stone.

## 5.2.6 Content Metadata from Video Repository

An incident of unauthorized deletion of content metadata stored in the video repository would cause a disruption of service. Once metadata is modified, the application will not be able to service requests from the middleware and content management applications.

Content metadata can also be modified or added via the administrative console, which could cause loss of integrity within the application. Modifications to the metadata will affect the integrity of the application and will require a detailed review of all metadata within the repository. Once integrity is affected, the IPTV service provider would be required to verify all entries on the system.

If the content metadata is easily accessible, an attacker could modify it to make the associated content widely available or reduce its cost. This would cause a loss in revenue to the IPTV service provider. It would also affect the content provider's perception of the service provider.

## 5.2.7 MPEG-2 Content from Video Repository

Any unauthorized deletion of MPEG-2 content stored in the video repository would cause a disruption of service. A similar effect can be caused by corrupting the contents or damaging the private keys used to recover the encrypted content.

Modifications of MPEG-2 content stored in the video repository via management console or by direct access to the repository would affect the integrity of the application. If files are modified or information inserted on MPEG-2 files, then the platform integrity is lost and administrators would be required to undertake a full review of all contents. There are malicious sequences of MPEG-2 frames that cause MPEG players to crash. If these frame sequences were inserted into a popular MPEG-2 movie, widespread service disruption would occur.

If the MPEG-2 content is stored in the clear, or if decryption keying material is accessible, an intruder could make a bit-by-bit copy of the unencrypted MPEG-2 content and distribute it to others. Internal threats are also present – employees at the IPTV service provider may also attempt to copy content without authorization.

## 5.2.8 MPEG-4 Content

MPEG-4 content can be modified or deleted by intruders or employees. These changes would cause service disruptions as the rest of the IPTV service would not be able to operate and fulfill user requests or provide broadcasts. Both the video repository application and the operating system are responsible for restricting access to the content.

If the MPEG-4 content is stored in the clear, or if decryption keying material is accessible, a thief could make a bit-by-bit copy of the unencrypted MPEG-4 content and distribute it to others, thereby depriving the content owner of revenue.

Unencrypted MPEG-4 content is also exposed to modifications or alterations of content. Any modification to the MPEG-4 content would damage the integrity of the repository and would cause delays in the service.

## 5.2.9 Load Balancer Software

Unauthorized access to the application may be used to stop the application or cause service disruptions such as those linked with changes on the parameters of the load balancer, removing elements from the list of supported servers or reducing the bandwidth allocation.

A traffic flood attack on the VOD VLAN directed towards a regional head end could consume all of the bandwidth on the in-bound access link of the regional head end, thus interrupting the VOD service originating from it. The same attack applies to the broadcast TV VLAN.

## 5.2.10 Master Video Streaming Software

The master video streaming server receives the connection management request from the IPTV middleware and redirects the subscriber's connection to the appropriate video cache streaming server or gaming server.

Buffer overflow vulnerabilities would be in the master video streaming server software itself and are caused by improper coding techniques and inadequate testing. These would be triggered by messages received over the middleware VLAN and could result in the overwriting of master video streaming server software or data.

Unauthorized access to the administrative console may be used to stop the master video streaming server software (e.g. a kill command). Likewise, an unauthorized command or protocol message that causes the master video streaming server software to halt itself (e.g. an 'exit' command) can be received over the middleware VLAN.

## 5.2.11 CA/DRM Service

Buffer overflow vulnerabilities could be exploited within the CA/DRM client software and used to modify the operation of the application. Intruders can use the existing vulnerabilities to capture private keys or plant spyware on the system. This type of attack would affect the integrity of the application and would leave the system exposed to future manipulation.

Unauthorized access to the management console or malformed requests over the network could be used to damage the operation of the system or stop the operation of the application. If the CA/DRM application is not operational, the rest of the IPTV service may be affected as encryption keys will not be available for set top boxes.

## 5.2.12 SRTP Keys

Unauthorized access to the administrative console may be used to delete the SRTP key. This type of attack can be undertaken by both internal and external attackers and would cause service disruption to IPTV subscribers. SRTP keys can be captured by intruders and used to create false keys or to have access to encrypted content. SRTP keying material could be captured in transit by means of a publicly available packet sniffer (e.g. ethereal) or MITM. These attacks would require access to the middleware VLAN between the CA/DRM system and the regional head-end egress point.

A forged decryption vector could be injected into the connection or a man-in-the-middle (MITM) attack could intercept and modify the decryption vector in transit with the result that the video cannot be decrypted. These attacks would require access to the middleware VLAN between the CA/DRM system and the regional head-end egress point.

## 5.2.13 Ismacryp Key

Unauthorized access to the administrative console can be used to delete the Ismacryp key. This will affect the operation of the service owing to the lack of keys. Similar effects would be caused if keys were modified by intruders or replaced in transit. This would cause service disruptions to subscribers.

A forged Ismacryp decryption vector or decryption keying material could be injected into the connection. An MITM attack could intercept and modify the Ismacryp decryption keying material or decryption vector in transit. The result of both of these attacks is that the video cannot be decrypted. For the current scope, these attacks would originate in the middleware VLAN between the CA/DRM system and the regional head-end egress point.

If the Ismacryp key or keying material is accessible, a thief could make a copy of the subscriber's Ismacryp key or keying material and use it to obtain content without paying for it. Attackers could try to guess and then obtain vulnerable Ismacryp keys, which would allow access to content without paying for it. If the Ismacryp key or keying material is accessible, an attacker could modify the key lifetime to obtain additional content. For example, in order to view the content for a longer time period, etc., without paying for it.

Packet flood of illegitimate key management message packets to the CA/DRM system could cause the CA/DRM system to stop responding, or a traffic flood on the middleware VLAN could consume all of the bandwidth of the connection, preventing legitimate traffic from getting through. These attacks would require access to the middleware VLAN between the CA/DRM system and the regional head-end egress point.

## 5.2.14 Key Management Protocol

Malformed key management protocol packets have the potential to result in a buffer overflow or cause the CA/DRM system to crash if improperly implemented. These packets could be forged or modified in transit.

Key management protocol packets could be captured in transit by means of a publicly available packet sniffer (e.g. ethereal) or MITM. These attacks would require access to the middleware VLAN between the CA/DRM system and the regional head-end egress point.

Malformed key management protocol packets have the potential to result in a buffer overflow or system crash if improperly implemented, potentially interrupting the VOD service.

An attacker could send a flood of key management protocol packets to the CA/DRM system in order to cause the server to crash. In addition, a traffic flood on the middleware VLAN would prevent legitimate traffic from getting through. These attacks could originate from the middleware VLAN between the CA/DRM system and the regional head-end egress point.

## 5.2.15 CA/DRM Service Administration

Unauthorized access to the administrative console can be used to send commands over the management network and delete the CA/DRM service administrative authentication information. Malformed packets or buffer overflows may be used to cause disruptions to the service.

Unauthorized duplication of CA/DRM service administrative authentication information could be the result of an intruder able to perform management activities. Authentication information can be later used for unauthorized access to the application.

## 5.2.16 VOD Application – Cached Video Content Metadata

Note that, since this information is cached, these threats can be rectified by recaching the information. The content distribution service is responsible for caching video content.

Unauthorized deletion of cached video content metadata stored in a video cache streaming server would be accomplished by unauthorized management activity or malware. Unauthorized access to the administrative console can be used to send commands over the management network and delete cached video content metadata stored in the video cache streaming server. The same attack can be used to modify cached copies of video content.

Unauthorized insertion or modification of cached video content metadata stored in a video cache streaming server would be accomplished by unauthorized management activity, malware or rootkit.

Unauthorized browsing can occur after a system break-in by an intruder (external attack) or by an insider gaining unauthorized access to the cached video content metadata. If this content is not encrypted, intruders would be able to retrieve copies for later use.

## 5.2.17 Cached MPEG-2/MPEG-4 Content (Primary and Secondary Storage)

Note that, since this information is cached, these threats can be rectified by recaching the information. The content distribution service is responsible for caching video content.

Unauthorized deletion of cached MPEG-2/MPEG-4 video content stored in the video cache streaming server would be accomplished by unauthorized management activity, malware or rootkit. Unauthorized access to the administrative console can be used to send commands over the management network that delete cached MPEG-2/MPEG-4 video content stored in the video cache streaming server.

Unauthorized insertion or modification of cached MPEG-2/MPEG-4 video content stored in the video cache streaming server would be accomplished by unauthorized management activity, malware or rootkit. Unauthorized access to the administrative console can be used to send commands over the management network and insert or modify MPEG-2/MPEG-4 content stored in the video cache streaming server. Intruders with access to the server may be able to replace or modify stored content, affecting the integrity of the platform.

Unauthorized browsing can occur after a system break-in by an intruder (external attack) or by an insider gaining unauthorized access to the cached MPEG-2 video content. If the cached MPEG-2/MPEG-4 video content is stored in the clear, or if decryption keying material is accessible, a thief could make a bit-by-bit copy of the unencrypted cached MPEG-2/MPEG-4 video content and distribute it to others, thereby depriving the content owner of revenue.

### 5.2.18 Video Streaming Software

Buffer overflow vulnerabilities would be in the video streaming software itself and are caused by improper coding techniques and inadequate testing. These would be triggered by messages received over the VOD VLAN or content distribution network and could result in the overwriting of video streaming software or data. Malformed MPEG-2 or MPEG-4 frames residing in the cached video content could also result in buffer overflow of the video streaming software.

Unauthorized access to the administrative console can be used to send commands over the management network that halt the video streaming software (e.g. a kill command). Likewise, an unauthorized command or protocol message that causes the video streaming software to halt itself (e.g. an 'exit' command) can be received over the VOD VLAN. Management commands may also be executed by an intruder able to perform management activities. This could also be accomplished by malware.

### 5.2.19 Local Ad Insertion Authentication Information (e.g. User ID(s) and Password(s))

Unauthorized access to the administrative console can be used to send commands over the management network that delete or modify the local ad insertion authentication information. Without valid authentication the application may not be able to communicate with other components and will not be operational.

Weak passwords may allow intruders easily to guess authentication information and have unauthorized access to the system. Unauthorized duplication of local ad insertion authentication information could be the result of an intruder able to perform management activities; it could also be the result of malware. Once duplicated, the authentication information can be taken off site to be decrypted, or in some cases replay attacks may be possible using previous authentication data.

Keyloggers and spyware could intercept the communications from the application administrator while he or she is entering authentication information to the local ad insertion system and send it back to the attacker. Shoulder surfing is another threat – visitors to the head end may try to view passwords while they are being typed.

An unauthorized management command can be used to remove authentication data, affecting the service. Additionally, intruders may cause accounts to lock out after a number of failed authentication attempts.

### 5.2.20 Local Ad Metadata

Unauthorized access to the administrative console can be used to send commands over the management network that delete or modify local ad content metadata stored in the local ad insertion system.

Unauthorized browsing can occur after a system break-in by an intruder (external attack) or by an insider gaining unauthorized access to the local ad content metadata. If the local ad content metadata is easily accessible, an attacker could modify it to make the associated local ad content available before launch of an advertising campaign, for example.

Unauthorized deletion or modification of local ad content metadata stored in the local ad insertion system could prevent access to the local ad content.

### 5.2.21 Local Ad MPEG-2/MPEG-4 Content

Unauthorized access to the administrative console can be used to send commands over the management network that delete local ad MPEG-2/MPEG-4 content stored in the local ad insertion system. Similar situations may allow intruders to insert or modify MPEG-2/MPEG-4 contents. Modified content may cause service disruptions or reputation damage to the IPTV service provider.

Unauthorized browsing can occur after a system break-in by an intruder (external attack) or by an insider gaining unauthorized access to the local ad MPEG-2/MPEG-4 content. If the local ad MPEG-2/MPEG-4 content is stored in the clear, or if decryption keying material is accessible, an intruder could make a bit-by-bit copy of the unencrypted local ad MPEG-2/MPEG-4 content and distribute it to others, potentially violating the copyright of the local ad.

### 5.2.22 Local Ad Insertion Tracking Information

Unauthorized access to the administrative console can be used to send commands over the management network that delete the local ad insertion tracking information. Unauthorized deletion of local ad insertion tracking information could also occur as a result of an internal or external attack or be caused by malware. A possible motive for deleting local ad insertion tracking information would be to reduce the amount the advertiser has to pay the IPTV service provider.

Unauthorized browsing can occur after a system break-in by an intruder (external attack) or by an insider gaining unauthorized access to the local ad insertion tracking information.

Unauthorized interception of local ad insertion information in transit from the regional head end to back-office systems could be accomplished by placing a packet sniffer device or using packet sniffing software or by an MITM attack.

One possible motive for duplicating the local ad insertion tracking information would be to sell this information to competitors. This information tends to match the demographics of subscribers as well as their purchasing behavior; competitors may want these valuable data to prepare their own campaigns.

### 5.2.23 nPVR Application Recorded/Stored Content Metadata

Unauthorized access to the administrative console can be used to send commands over the management network that delete/insert recorded video content metadata stored in the nPVR application. This management activity could also be the result of an external or internal attack.

Unauthorized browsing can occur after a system break-in by an intruder (external attack) or by an insider gaining unauthorized access to the recorded/stored video content metadata.

Unauthorized deletion or modification of recorded/stored video content metadata stored in the nPVR application could prevent playback of recorded/stored video content.

### 5.2.24 Recorded/Stored MPEG-2/MPEG-4 Content

Unauthorized access to the administrative console can be used to send commands over the management network that delete recorded/stored MPEG-2/MPEG-4 content stored in the nPVR application.

Unauthorized access to the administrative console can be used to send commands over the management network that insert or modify recorded/stored MPEG-2/MPEG-4 content stored in the nPVR application.

If the recorded/stored MPEG-2/MPEG-4 content is stored in the clear, or if decryption keying material is accessible, an intruder could make a bit-by-bit copy of the unencrypted recorded/stored MPEG-2/MPEG-4 content and distribute it to others, resulting in loss of revenue to the content owner.

Unauthorized browsing can occur after a system break-in by an intruder (external attack) or by an insider gaining unauthorized access to the recorded/stored MPEG-2/MPEG-4 content.

### 5.2.25 nPVR/Video Recording Software

Buffer overflow vulnerabilities could be found on the nPVR/video recording software itself and are caused by improper coding techniques and inadequate testing. These would be triggered by instructions received over the VOD VLAN and could result in the overwriting of nPVR/video recording software or data.

Unauthorized access to the administrative console can be used to send commands over the management network that halt the nPVR/video recording software (e.g. a kill command). Likewise, an unauthorized command or protocol message that causes the nPVR/video recording software to halt itself (e.g. an 'exit' command) can be received over the VOD VLAN. Management commands may also be executed by an intruder able to perform management activities. This could also be accomplished by malware.

Overwriting of nPVR application software or data, as well as malformed packets received over the VOD VLAN, could cause the software to hang, crash or terminate prematurely.

## 5.3 IPTV Network Provider – Transport and Aggregation Network

### 5.3.1 Protocol Vulnerabilities

There are several protocols involved in the IPTV operations, each one with different security problems and vulnerabilities.

**Multicast Vulnerabilities (IGMP, Multicast Source Discovery Protocol Vulnerabilities)**
As a multicast protocol, any vulnerability with IGMP can be significantly greater than for other protocols. This is because of the impact that exploitation of IGMP vulnerabilities could

have. With a single attack, several hundreds or even thousands of hosts (subscribers, set top boxes) could suffer.

Among known multicast security problems, vulnerability of the infrastructure to DOS attacks is of prime concern in order to ensure IPTV service availability. Content is protected by DRM, and it would be difficult to extract the video without proper keys. The broadcast nature of multicast magnifies the effects of DOS attacks by the number of subscribers. A secondary effect is that other services may be affected. In the case of 'triple-play' users, other services could suffer a knockdown effect. If quality of service and bandwidth management mechanisms have not been deployed, attackers using vulnerabilities or design problems on the multicast network could take down the high-speed Internet access, VoIP and IPTV VLANs.

One example of this situation is the vulnerabilities found on the multicast source discovery protocol (MSDP) – RFC 3618. A significant part of the Internet infrastructure relies on a model called 'Any Source Multicast' (ASM). ASM is supported by four protocols: IGMP, the protocol independent multicast (PIM), the multicast border gateway protocol (MBGP) and MSDP. MBGP supplies interdomain route exchange information that allows the dissemination of reach ability and path information across domains. PIM uses the information to manage distribution trees that are transferred to its participants. Hosts begin group adhere and leave requests using IGMP. The MSDP is used as a source announcement protocol and propagates information about currently active sources.

Once the service provider starts broadcasting to a group, the first hop router (FHR) creates a PIM register message that links the source–group pair. The PIM message is then sent to the PIM rendezvous point (RP) which acts as an MSDP router for the domain. Then, using MSDP source active (SA) messages, the information regarding the source is propagated. If an STB or cable modem needs to receive the content, it sends an IGMP join request to its FHR. Then the FHR of the STB sends a PIM join message to the RP in its domain. Consequently, the PIM distribution tree is extended to the receiver so that it receives data from the service provider.

The MSDP is used to maintain a current list of the IP addresses used by the different sources within the infrastructure.

There are several security issues with the way multicast operates: confidentiality, access control and control protocol traffic.

By design, multicast traffic is intended to be received by multiple recipients. Encrypting the content for an individual recipient is highly complex, and this is left to the DRM software. Any information distributed by the multicast protocols will not be protected by the protocol and will rely on additional protection being provided by underlying applications.

Regarding access controls, there are not many resources available. Set top boxes are allowed to join groups without much restriction. Multicast protocols do not provide much access control, relying on additional functions at DSLAMs and edge gateways to authenticate and validate users. Once joined, hackers could send a high volume of traffic to a multicast group and cause a DOS attack.

Another type of attack would be to send a broadcast message claiming that there is a high loss rate on the channel or high congestion, and the source would reduce the transmission rate, affecting other subscribers.

Malicious control traffic can also be injected into the communication, causing applications to malfunction, or traffic to be sent to the wrong destination. IGMP traffic can be easily spoofed owing to the lack of authentication fields within the stream. An additional effect of this vulnerability is the risk of state information being modified by intruders.

Forged control messages can be used to alter the structure of multicast distribution trees and affect the data distribution across them. The vulnerabilities in MBGP are due to the security problems of the underlying BGP protocol.

Vulnerabilities in MSDP arise owing to security problems with control traffic (source active – SA), as it can easily be spoofed and sent throughout the infrastructure. An intruder can create a large number of SAs, causing all MSDP routers in the infrastructure to have incorrect state data.

### Vulnerabilities of MSDP to DOS Attacks

As IPTV infrastructures rely on multicast and MSDP to operate, intruders could use vulnerabilities in the underlying protocol to cause the whole infrastructure to collapse owing to excess load.

MSDP DOS attacks attempt to flood the network with falsified source active (SA) messages. As part of MSDP, when a source in a multicast network first sends an IP packet to a new class D address, an MSDP peer in the domain sends a corresponding SA announcement to all of its peers. Also, every MSDP peer broadcasts all SAs it receives to every other connected MSDP peer, thus creating the risk of flooding.

### Networking Threats

#### (i) Denial of Service (DOS)

IPTV is expected to be an always-on-stream service. Subscribers are used to the satellite or cable service and will not accept an intermittent service. A bigger impact would be suffered if the VOD service were affected, as the content provider would lose important revenues from the VOD sales.

These factors support the idea that denial of service is one of the biggest threats to the IPTV market, and understanding the technology that can be used to secure content delivery shows that hackers could be left with only one option to attack IPTV networks. As confidentiality and integrity are ensured by the use of encryption, only the availability can be attacked.

#### (ii) Distributed Denial of Service

A D-DOS is basically a series of servers and desktops that have been infected by a virus or Trojan horse and are manipulated by a hacker. The infected systems are called 'zombies' or 'bots'. The hacker then instructs the zombies to send a very large number of requests to a particular host, crashing the network or the server.

There are many reported cases where criminal gangs have blackmailed websites and demanded payments so a D-DOS is not sent against the system.

The impact of a D-DOS on an IPTV network should be considered when defining the architecture. This is particularly true if the hackers use zombies from within the service provider clients. A D-DOS of cable modem users from within the same network will crash the service.

#### (iii) Smurf Attack

A smurf attack is a variation of a DOS attack using the broadcast functionality of TCP/IP. It works by sending an ICMP echo (ping) to an Internet broadcast address using a spoofed address. This broadcasts the message to all hosts connected to the subnet, and all computers in the subnet reply to the spoofed address. All other hosts receiving the reply to the spoofed broadcast address will reply and a large amount of traffic will be created.

### UDP Flood
User datagram protocol flood can be undertaken using vulnerabilities in the echo and chargen services. Intruders trick the service into answering spoofed requests or creating an infinite loop that uses all the system resources and crashes the system.

### Fragmentation Attack
This type of attack involves sending a series of packets with apparent nondangerous load in order to sneak past firewalls or intrusion detection systems. When the receiving system reassembles the packets, a malformed packet or system attack will be revealed and in some cases the server will crash.

### Syn Flood
During the initial TCP/IP session handshake, several packets are sent to and from both elements involved. In some cases, if one of the two is not answering, the other will keep a certain amount of memory allocated to answer once the packet is received. Over time, the system may expend all available memory and crash.

Most current operating systems have basic countermeasures against this type of attack. but some embedded systems with primitive TCP/IP stacks may fall victim to syn flood attacks.

## 5.3.2 Content Distribution Service: Unicast Content Propagation – FTP or Other Transport Protocol

Forged or unauthorized modification of FTP (or other transport protocol) packets in transit would most likely result in the loss of fidelity of the video stream or the insertion of bogus content into the video stream upon playback by the video cache streaming server located in the regional head end. Bogus content can also be inserted into the application (game) being distributed to gaming servers via the content distribution network. This could be accomplished by injecting forged packets into the transport stream or by an MITM attack.

The FTP (or other transport protocol) packets could be captured in transit by means of a publicly available packet sniffer (e.g. ethereal) or MITM. Once these packets have been captured, they can be played back at a later time conceivably to a recording device for personal use or distribution.

An attacker could send a flood of FTP (or other transport protocol) packets to the video cache streaming servers or gaming servers in order to cause the servers to crash. A traffic flood to the content distribution network could consume all of the bandwidth on the connections between the content management system and the video cache streaming servers. This attack could originate from within the content distribution network between the centralized content management system and the video cache streaming servers and gaming servers located in the regional head ends.

## 5.3.3 Multicast Content Propagation

### 5.3.3.1 IGMPv2/v3 (Snooping)

Modifications by an attacker (MITM) to IGMP packets on the content distribution network in transit between content management system and video cache streaming servers and gaming servers in the regional head ends.

Malformed IGMP packets could be created by an attacker and sent to the content management system, video cache streaming servers or gaming servers. Malformed packets could cause buffer overflow or the system to hang/crash.

Forged IGMP packets could be created by an attacker and sent to the content management system, video cache streaming servers or gaming servers via the content distribution network. A replay attack could result in unauthorized participation in a multicast flow.

Forged IGMP packets created by an attacker could allow unauthorized participation in a multicast group. For example, forged join messages could result in the intruder participating in the content distribution group, resulting in the acquisition of content without paying for it. Forged leave messages can cause the generation of numerous bogus query messages from the multicast router.

Forged IGMP packets could be created by an attacker masquerading as a multicast router in order to acquire content.

There is a risk of intruders being able to undertake packet sniffing by deploying a protocol analyzer or using hosts on the IP subnet. Publicly available sniffing tools (e.g. ethereal) or an MITM on the content distribution network can discover/obtain content being distributed over the content distribution network. Observed IGMP packets provide information on what multicasts are active and the devices that are participating. These attacks would require access to the content distribution network between the centralized content management system and the video cache streaming servers and gaming servers located in the regional head ends.

Malformed IGMP packets received by the content management system or video cache streaming servers could cause buffer overflow or cause the system/server to hang or crash.

Since the multicast distribution trees are built on the basis of IGMP control messages, forged control packets could result in suppression or deflection of packet flows.

A packet flood of IGMP packets could result in a system/server panic, which might cause it to hang/crash. A traffic flood on the content distribution network could consume all the bandwidth on the network, preventing legitimate traffic from getting through.

IGMPv2 allows local users to cause a denial of service via an IGMP membership report to the Ethernet address of a target instead of the multicast group address, causing the target to stop sending reports to the router and effectively disconnecting the group from the network.

### 5.3.3.2 PIM (SM, SSM, Snooping)

There are threats of modifications by an attacker (MITM) to PIM packets in transit between the content management system and the video cache streaming servers and gaming servers located in the regional head ends.

Malformed PIM packets could be created by an attacker and sent to the content management system, video cache streaming servers or gaming servers. Malformed packets could cause buffer overflow or cause the system/server to hang/crash.

A malicious host or router can easily forge control messages, resulting in corrupted multicast distribution trees since they are built on the control messages.

Forged PIM packets created by an attacker could allow for unauthorized participation in a multicast group. Forged join messages could result in the intruder participating in the content distribution group, resulting in the acquisition of content without paying for it.

Observed PIM packets provide information on what multicasts are active and the devices that are participating. These attacks would require access to the content distribution network between the centralized content management system and the video cache streaming servers located in the regional head ends.

Malformed PIM packets received by the content management system, video cache streaming servers and gaming servers could cause buffer overflow or cause the system/server to hang or crash. Since the multicast distribution trees are built on the basis of PIM control messages, forged control packets would result in suppression or deflection of packet flows.

A packet flood of PIM packets could result in system/server panic, which might cause it to hang crash. A traffic flood on the content distribution network could consume all the bandwidth on the network, preventing legitimate traffic from getting through.

### 5.3.3.3 MBGP

MBGP adds features to BGP to enable multicast routing between BGP autonomous systems. MBGP would not be used for the content distribution network unless the content distribution network consisted of multiple autonomous systems. Since the content distribution network connects regional head ends and the content management system located in the national head end, it is conceivable that the content distribution network is composed of multiple autonomous systems. Vulnerabilities in MBGP are mainly due to the susceptibility of the underlying BGP protocol.

Malformed MBGP packets could be created by an attacker and sent to MBGP routers in the national head end, regional head ends or content distribution network. Malformed MBGP packets could cause a buffer overflow or cause the MBGP router to hang/crash.

Forged MBGP packets could be created by an attacker and sent to the MBGP routers in the national head end, regional head ends or content distribution network.

BGP has no internal mechanism to provide strong protection of the integrity, freshness and peer entity authenticity of the messages in peer-to-peer BGP communications. Additionally, there is no mechanism within BGP to validate the authority of an AS to announce NLRI information, and no mechanism within BGP to ensure the authenticity of the path attributes announced by an AS. Forged or modified MBGP packets could result in unauthorized participation in a multicast group. The intruder would participate in the content distribution group, resulting in the acquisition of content without paying for it.

A packet flood of MBGP packets could cause the MBGP router to crash or hang. A traffic flood on the content distribution network could consume all the bandwidth of the network, preventing legitimate traffic from getting through.

Malformed MBGP packets received by the MBGP router could cause the MBGP router to stop responding. Forged MBGP packets could corrupt the MBGP router's routing table, causing the MBGP router to suppress or deflect the distribution of content.

### 5.3.3.4 MSDP

MSDP SA messages contain information about multicast sources in PIM-SM domains.

Control traffic, for example source active (SA) messages, can easily be forged or modified and sent throughout the content distribution network in order to cause participation in a bogus multicast group. A motive for this would be to substitute illegitimate content for legitimate content. For the current scope, this attack would originate in the content distribution network between the centralized content management system and the video cache streaming servers located in the regional head ends.

There are no known vulnerabilities involving malformed MSDP packets.

Forged or modified MSDP SA messages created by rogue senders or MITM attacks can be used to allow for unauthorized participation in a bogus multicast group for the purpose of acquiring content.

System panic can be caused by a packet flood of MSDP packets (e.g. MSDP storms due to worm infection of the host). MSDP storms are the most common attacks seen on multicast networks to date (e.g. Ramen, Slammer/Sapphire, Sasser). They could cause the multicast routers located in the national or regional head ends to run out of memory and crash.

Because MSDP is responsible for propagating information about all active sources, a bogus MSDP peer can generate a large number of bogus source active (SA) messages and send these to every single other MSDP peer. This kind of simple attack can have widespread negative consequences, resulting in DOS.

Malformed MSDP packets received by multicast routers could cause buffer overflow or cause the multicast router to stop responding. A traffic flood on the content distribution network could consume all of the bandwidth on the network, preventing legitimate traffic from getting through.

Forged or modified MSDP packets by an attacker could cause the content management system, video cache streaming server(s) or gaming server(s) to join the wrong group, which would result in the suppression or deflection of traffic.


### 5.3.3.5 MFTP

Forged MFTP (or other transport protocol) packets or unauthorized modification of these packets in transit would most likely result in the loss of fidelity of the video stream or the insertion of bogus content into the video or application content being distributed. This could be accomplished by injecting forged packets into the content being transported or by an MITM attack.

MFTP (or other transport protocol) packets could be captured in transit by means of a publicly available packet sniffer (e.g. ethereal) or MITM. Once these packets have been captured, they can be played back at a later time conceivably to a recording device for personal use or distribution.

MFTP (or other transport protocol) packets could be captured in transit by means of a publicly available packet sniffer (e.g. ethereal) or MITM. Once these packets have been captured, they can be played back at a later time.

An attacker could send a flood of MFTP (or other transport protocol) packets to the video cache streaming servers or game servers in order to cause the servers to crash. A traffic flood on the content distribution network could consume all of the bandwidth on the network and prevent legitimate traffic from getting through. These attacks could originate on the content distribution network between the centralized content management system and the video cache streaming servers and gaming servers located in the regional head ends.

Altered, malformed or forged transport packets have the potential to result in a system crash, loss of fidelity of video playback or insertion of bogus video and application content, all of which translate into an interruption of IPTV service.

### 5.3.3.6 RTP

Forged RTP packets or unauthorized modification of RTP packets in transit would most likely result in the loss of fidelity of the video stream or the insertion of bogus content into the video or application content being transported. This could be accomplished by injecting forged packets into the RTP stream or by an MITM attack. For example, Ohrwurm is a small and simple RTP fuzzer. Some features include the ability to read SIP messages to get information about the RTP port numbers. Fuzzing of RTP traffic allows for MITM attacks, and the RTP payload is fuzzed with a constant BER. The BER is also configurable. For the current scope, these attacks would originate in the content distribution network between the centralized content management system and the video cache streaming servers and gaming servers located in the regional head ends.

A malicious sequence of RTP packets has been known to cause at least one RTP client to hang or crash.

RTP packets could be captured in transit by means of a publicly available packet sniffer (e.g. ethereal or rtpdump/rtpundump) or MITM. Once these packets have been captured, they can be played back at a later time conceivably to a recording device for personal use or distribution.

An attacker could send a flood of RTP packets to the video cache streaming servers or gaming servers in order to cause the servers to crash. A traffic flood on the content management network could consume all of the bandwidth on the network. This attack could originate on the content management network between the centralized content management system and the video cache streaming servers and gaming servers located in the regional head ends.

As described for the corruption threat above, altered, malformed or forged RTP packets have the potential to result in a system crash, loss of fidelity of video playback or insertion of bogus video or application content, all of which translate into an interruption of IPTV service.

## 5.3.4 QoS Signaling (RSVP, DiffServ)

There is a threat of suffering from a compromise of platform integrity.

Forged or modified (e.g. MITM) RSVP messages could be created by an attacker and sent to the content management system, video cache streaming servers or gaming servers. These messages could result in setting up an unauthorized QoS flow through the content distribution network or adjusting an existing QoS flow.

Corrupted routers may invalidly modify the contents of the RSVP messages and send them to the content management system, video cache streaming servers or gaming servers, or someone can just inject malformed RSVP messages into the network destined for these systems/servers. The malformed RSVP messages could cause the system/server to hang or crash.

There is a threat of modification of the DSCP field in the IP packet header in order to flood the particular DSCP or adjust the QoS associated with this flow in an unauthorized manner. For the current scope the modification could be done by the content management system, video cache streaming server(s) or gaming server(s).

The attacker can potentially intercept or drop all or some of the reservation messages such that the QoS reservation and channel set-up will fail or be maliciously delayed in a persistent way and cause a premature teardown. An attacker could also send forged RSVP messages that would tear down the channel.

A malformed RVSP packet may cause the content management system, video cache streaming server or gaming server to hang or crash, for example owing to buffer overflow.

A packet flood of RSVP packets to the content management system, video cache streaming server or gaming server could cause the device to hang or crash owing to system overload.

A traffic flood on the content distribution network could consume all of the bandwidth on the network.

An attacker may request resource reservations for any amount of bandwidth and produce a corresponding amount of traffic on the content distribution network, thereby using all of the bandwidth on the netowork, preventing legitimate traffic from getting through.

An attacker could set the DSCP field to a particular value in the IP header of every packet in order to flood the QoS flow.

## 5.3.5 Management of Content Distribution Service

There is a threat of unauthorized management activity resulting in the deletion of content distribution service administrative authentication information. Unauthorized access to the administrative console can be used to send commands over the management network that delete the content distribution service administrative authentication information.

There is a threat of unauthorized management activity resulting in tampering with (or modification of) content distribution service administrative authentication information. Unauthorized access to the administrative console can be used to send commands over the management network that modify the content distribution service administrative authentication information. Unauthorized duplication of content distribution service administrative authentication information could be the result of an intruder able to perform management activities. An intrusion may be the result of an external or internal attack.

An unauthorized administrative command or protocol message can be received over the content distribution network that duplicates content.

If an attacker is able physically to observe the content distribution service administrator entering his or her authentication information, it can be reused at a later time by the attacker to access the content management system.

If the content distribution service administrator authentication information is deleted, the content management system will not allow the administrator to log in.

An attacker could deny the content distribution service administrator access to the content management system by knowingly attempting to log in with improper authentication information until the account lockout threshold is exceeded.

Unauthorized deletion or modification of content or content metadata could prevent playback of the content.

## 5.3.6 Connection Management Service

The connection management service establishes a connection between a subscriber and a video cache streaming server or gaming server located in a regional head end. The connection management service is invoked by the IPTV middleware upon receipt of a subscriber request.

### 5.3.6.1 DSM-CC Protocol

DSM-CC commands are forwarded from the middleware server to the master video streaming server in order to direct it to establish a connection between the subscriber and video cache streaming server to stream VOD content to the subscriber.

Malformed DSM-CC packets have the potential to result in a buffer overflow or cause the master video streaming server to crash if improperly implemented. These packets could be forged or modified in transit. Malformed DSM-CC packets have the potential to result in a buffer overflow or system crash if improperly implemented, potentially interrupting the VOD service.

DSM-CC packets could be captured in transit by means of a publicly available packet sniffer (e.g. ethereal) or MITM.

As described for the corruption threat above, malformed DSM-CC packets have the potential to result in a buffer overflow or system crash if improperly implemented, potentially interrupting the VOD service.

An attacker could send a flood of DSM-CC packets to the master video streaming server in order to cause the server to crash. This attack could originate from the IP subnet containing the middleware server and master video streaming server.

A traffic flood on the IP subnet containing the middleware server and master video streaming server could consume all of the bandwidth on the network and prevent legitimate traffic from getting through.

### 5.3.6.2 RTSP Protocol

RTSP commands are forwarded from the middleware server to the master video streaming server in order to direct it to establish a connection between the subscriber and video cache streaming server to stream VOD content to the subscriber.

Malformed RTSP packets have the potential to result in a buffer overflow or cause the master video streaming server to crash if improperly implemented. These packets could be forged or modified in transit. Malformed RTSP packets have the potential to result in a buffer overflow or system crash if improperly implemented, potentially interrupting the VOD service.

RTSP packets could be captured in transit by means of a publicly available packet sniffer (e.g. ethereal) or MITM. For the current scope, this attack would require access to the IP subnet containing the middleware server and master video streaming server.

An attacker could send a flood of RTSP packets to the master video streaming server in order to cause the server to crash. In addition, a traffic flood on the IP subnet containing the middleware server and master video streaming server would prevent legitimate traffic from getting through. These attacks could originate from the IP subnet containing the middleware server and master video streaming server.

### 5.3.6.3 MPEG-2 Video Stream

Malformed MPEG-2 packets have the potential to result in a buffer overflow or cause the system to crash if improperly implemented. A specially crafted MPEG-2 stream can cause a buffer overflow in vulnerable software, allowing remote attackers to execute arbitrary code. Multiple multimedia players (e.g. Quicktime, xine) are susceptible to these types of attack.

Forged MPEG-2 packets or unauthorized modification of MPEG-2 packets in transit would most likely result in the loss of fidelity of the MPEG-2 stream. This could be accomplished by injecting noise into the MPEG-2 stream or by an MITM attack.

MPEG-2 packets could be captured in transit by means of a packet sniffer (e.g. dvbsnoop) or MITM, resulting in access to content without paying for it. This content could subsequently be redistributed. These attacks would require access to the VOD VLAN between the video cache streaming server and the regional head-end egress point.

Encrypted MPEG-2 packets could be captured in transit by means of a publicly available packet sniffer (e.g. ethereal) or MITM. The attacker could then perform decryption attacks against these captured packets (e.g. brute-force attack against the decryption key) offline. For the current scope, this attack would require access to the VOD VLAN between the video cache streaming server and the regional head-end egress point.

An attacker could send a flood of MPEG-2 packets to the STB in order to cause the STB to crash. Owing to the limited bandwidth of the STB interfaces, this could be a DOS (as opposed to a D-DOS). In addition, a traffic flood on the VOD connection could consume all of its bandwidth and prevent legitimate traffic from getting through. For the current scope, these attacks could originate from the VOD VLAN between the video cache streaming server and the regional head-end egress point.

### 5.3.6.4 MPEG-4 Video Stream

Malformed MPEG-4 packets have the potential to result in a buffer overflow or cause the system to crash if improperly implemented.

Forged MPEG-4 packets or unauthorized modification of MPEG-4 packets in transit would most likely result in the loss of fidelity of the MPEG-4 stream. This could be accomplished by injecting noise into the MPEG-4 stream or by an MITM attack.

MPEG-4 packets could be captured in transit by means of a packet sniffer (e.g. dvbsnoop) or MITM, resulting in access to content without paying for it. This content could subsequently be redistributed. These attacks would require access to the VOD VLAN between the video cache streaming server and the regional head-end egress point.

Encrypted MPEG-4 packets could be captured in transit by means of a publicly available packet sniffer (e.g. ethereal) or MITM. The attacker could then perform decryption attacks against these captured packets (e.g. brute-force attack against the decryption key) offline. This attack would require access to the VOD VLAN between the video cache streaming server and the regional head-end egress point.

An attacker could send a flood of MPEG-4 packets to the STB in order to cause the STB to crash. Owing to the limited bandwidth of the STB interfaces, this could be a DOS (as opposed to a D-DOS). In addition, a traffic flood on the VOD connection could consume all of its bandwidth and prevent legitimate traffic from getting through. These attacks could originate from the VOD VLAN between the video cache streaming server and the regional head-end egress point.

### 5.3.6.5 DSM-CC

DSM-CC is used to provide VCR-like controls for MPEG-2 flows. The security vulnerabilities of these types of protocol have not been fully explored.

Malformed DSM-CC packets have the potential to result in a buffer overflow or cause the video server to crash if improperly implemented, potentially interrupting the VOD service. The direction of the attack would be towards the video server and away from the STB.

Forged DSM-CC packets or unauthorized modification of DSM-CC packets in transit would most likely result in the transmission of bogus trick play commands to the video server. These would be nuisance-type attacks (e.g. bogus STOP commands) that could interrupt VOD service. The direction of attack would be towards the video server and away from the STB. This could be accomplished by injecting packets into the DSM-CC connection or by an MITM attack. This attack would occur on the VOD VLAN between the regional head-end access point and video cache streaming server.

An attacker could send a flood of DSM-CC packets to the video server in order to cause the video server to crash. For the current scope, this attack would occur on the VOD VLAN between the regional head-end access point and video cache streaming server.

A traffic flood on the VOD VLAN could consume all of its bandwidth, potentially preventing legitimate traffic from getting through.

### 5.3.6.6 RTSP

RTSP is used to provide VCR-like controls to MPEG-4 flows.

Malformed RTSP packets have the potential to result in a buffer overflow or cause the system to crash if improperly implemented, potentially interrupting the VOD service. Multiple multimedia players (e.g. Quicktime, RealNetworks) are susceptible to these types of attack.

Forged RTSP packets or unauthorized modification of RTSP packets in transit would most likely result in bogus trick play commands being sent to the video server on behalf of the subscriber. These would be nuisance-type attack (e.g. bogus STOP commands) that could interrupt VOD service. This could be accomplished by injecting packets into the RTSP connection or by an MITM attack. These attacks would occur on the VOD VLAN between the regional head-end access point and video cache streaming server.

An attacker could send a flood of RTSP packets to the video server in order to cause the video server to crash. In addition, a traffic flood on the VOD VLAN would prevent legitimate traffic from getting through to the video server. These attacks would occur on the VOD VLAN between the regional head-end access point and video cache streaming server.

## 5.4  IPTV Subscriber – Home End

### 5.4.1  Set Top Box

There are many different set top box models available on the market. Some are manufactured using both proprietary hardware and software. Others are based on open-source operating systems. Recently, some have started to use standard PC technologies on small-form hardware, allowing subscribers to load their own OS, middleware and DRM clients.

Some examples of operating systems used on set top boxes are as follows.

*Windows CE*

This is just an example of the several embedded operating systems that would be found on an IPTV infrastructure. Windows CE was designed as a very secure 'closed' operating system that blocked access to the kernel.

In September 2004, a hacker known as 'Ratter' discovered a vulnerability with this system, allowing him to create the first reported virus on a Windows CE system.

This information disclosure vulnerability is reported to affect the Windows CE kernel.

It is reported that the kernel memory structure KDataStruct is available to userland applications. Ultimately, this can be employed on any Windows CE system to gain addresses of the export sections of several kernel libraries.

This vulnerability is exploited by the virus WinCE.Duts.A (MCID 3238) in order to provide portability and reliability.

The fact that a virus was created for this platform shows that, in the future, more security problems will be found, and it is important to maintain a constant review of the behavior of STB to spot security manipulation.

*Set Top Boxes*

Set top boxes (STBs) have been used by the cable industry for several years, and their purpose is to decode digital signals into analog signals for TV sets. Over time, their capabilities and flexibility have grown, and now many companies use digital STBs with special features.

IPTV also requires STBs to decode digital information as TVs lack the necessary tuner to decode the signals. Additionally, IPTV two-way functions require more processing power and storage that must be provided by external equipment until TV set manufacturers start adding these capabilities to their products.

Set top boxes include basic security levels managed by content owners. These include the capability to authorize the type of content that subscribers are entitled to see. Most conditional access systems are supported by a smart card to store authentication information about the subscriber. These smart card functions are starting to be expanded by using the flexibility and strength of PKI systems supported by the ITU-T X.509.

X.509 defines the structure, fields and overall standard for digital certificates and digital signatures. By using the X.509 certificates incorporated as part of the STB content, providers will be able to apply security measures on a per subscriber basis. Requests could be digitally signed using the subscriber's digital certificate, creating nonrepudiation for service request transactions. There is an inherent risk to this technology, as hackers will eventually take control of the boxes and, capturing the certificate, will be able to order contents and even broadcast the stream to other subscribers. Worse still, if hackers capture the digital certificate from the content provider, they will be able to modify the stream and include any information they want.

For decades embedded system designers have been shielded from these security issues because their systems operated on closed secure environments and there was no incentive or possibility to create a self-replicated worm that crashed embedded systems. This situation might change in the future. With more and more cable modems and set top boxes, hackers will eventually target these systems in the same way as mobile phones have been victims of the first generation of viruses for mobile devices.

Games and storage functions are gradually being added to set top boxes. The operating system and software required to support such applications will allow an easier target for hackers and virus writers, as they are already familiar with such operating systems and viruses and vulnerabilities have been found on these systems in the past.

Satellite TV operators have the longest history with security incidents involving smart cards and encryption. The system is based on a combination of a generic STB and a smart card issued specifically to the subscriber that contains personal information and the information to recover the encrypted satellite signal. For several years, hackers were able to copy the originals and redistribute cloned cards for unlimited access to the satellite service. This caused millions of dollars in losses to the industry and was followed by a vicious circle of small changes to the security and immediate reaction by hackers. Apparently, each time the security was improved, hackers were able to circumvent it in a few hours or days, reissuing cards and continuing with the fraud.

It took a lot of effort and a series of small updates until the satellite providers were able to reprogram their cards to look for evidences of fraud and disable themselves. This case shows how set top boxes can be manipulated by hackers to gain free access to contents and how important it is for content providers to maintain the highest security levels from inception and not to take a reactive stance.

## 5.4.2 STB Executing Software

### 5.4.2.1 DRM Software

Buffer overflow vulnerabilities would be in the DRM client software itself and are caused by improper coding techniques and inadequate testing. These would be triggered by messages received over the middleware VLAN or home network (e.g. insecure WLAN) and could result in the overwriting of DRM software or data.

*Service Disruption*
Unauthorized access to the administrative console can be used to send commands over the management VLAN or home network (e.g. insecure WLAN) that halt the DRM software (e.g. a kill command). Likewise, an unauthorized command or protocol message that causes the DRM software to halt itself (e.g. an 'exit' command) can be received over the middleware VLAN or home network. Management commands may also be executed by an intruder able to perform management activities. This could also be accomplished by malware.

Overwriting of DRM software or data via a buffer overflow could cause the DRM software to hang, crash or terminate prematurely.

### 5.4.2.2 Middleware Client SW

Buffer overflow vulnerabilities would be in the middleware client software itself and are caused by improper coding techniques and inadequate testing. These would be triggered by messages received over the middleware VLAN or home network (e.g. insecure WLAN) and could result in the overwriting of middleware client software or data.

A forged user command would be sent over the middleware VLAN towards the middleware server. This user command could originate from the home environment (e.g. compromised WLAN) or be injected into the middleware VLAN, residential gateway or DSLAM.

A forged service response would be sent to the Middleware client in the subscriber's STB. This service response could originate from the home environment (e.g. compromised WLAN) or be injected into the middleware VLAN, residential gateway or DSLAM.

User interactions can be intercepted by sniffing middleware VLAN or by installing spyware on the STB (including a corrupted copy of the middleware client software itself, e.g. rootkit). The sniffing can be accomplished using a publicly available packet sniffer (e.g. ethereal) or MITM and can occur on the transmission facilities or from within a compromised DSLAM or residential gateway. These attacks would require access to the home network environment or the middleware VLAN between the DSLAM and the residential gateway.

Publicly available tools can be used to discover the existence of middleware client software executing on the STB and potentially its vendor and version number, which can be used to search for known vulnerabilities that can be exploited.

Unauthorized access to the administrative console can be used to send commands over the management VLAN or compromised home network that halt the middleware client software (e.g. a kill command). Likewise, an unauthorized command or protocol message that causes the middleware client software to halt itself (e.g. an 'exit' command) can be received over the middleware VLAN or compromised home network.

Overwriting of middleware client software or data via a buffer overflow could cause the middleware client software to hang, crash or terminate prematurely.

Malformed messages received over the middleware VLAN or compromised home network destined for the middleware client software could cause the middleware client software to hang, crash or terminate prematurely.

### 5.4.2.3 STB Platform SW

Buffer overflow vulnerabilities would be in the STB platform software (e.g. LINUX, MS-TV) itself and are caused by improper coding techniques and inadequate testing. These would be triggered by messages received over the WAN connection or home network (e.g. insecure WLAN) and could result in the overwriting of STB platform software or data.

User interactions can be intercepted by installing spyware on the STB (including a corrupted copy of the STB platform software itself, e.g. rootkit).

Publicly available tools can be used to discover the existence of STB platform software executing on the STB and potentially its vendor and version number, which can be used to search for known vulnerabilities that can be exploited.

An unauthorized management command or protocol message (e.g. SNMP) can be received over the management VLAN or compromised home network (e.g. insecure WLAN) that halt the STB platform software (e.g. a reboot command). Management commands may also be executed by an intruder who is able to perform management activities. This could also be accomplished by malware.

Overwriting of STB platform software or data via a buffer overflow could cause the STB platform software to hang, crash or terminate prematurely.

### 5.4.2.4 DVR/PVR

Buffer overflow vulnerabilities would be in the DVR/PVR software itself and are caused by improper coding techniques and inadequate testing. These would be triggered by messages received over the WAN connection or home network (e.g. insecure WLAN) and could result in the overwriting of DVR/PVR software or data.

User interactions with the DVR/PVR can be intercepted by sniffing the subnet that carries the user's middleware VLAN or by installing spyware on the STB (including a corrupted copy of the STB-based DVR/PVR software itself, e.g. rootkit). The sniffing can be accomplished using a publicly available packet sniffer (e.g. ethereal) or MITM and can occur on the transmission facilities or from within a compromised DSLAM or residential gateway. These attacks would require access to the home network environment or the middleware VLAN between the DSLAM and the residential gateway.

Publicly available tools can be used to discover the existence of executing STB-based DVR/PVR software and potentially its vendor and version number, which can be used to search for known vulnerabilities that can be exploited.

Unauthorized access to the administrative console can be used to send commands over the management VLAN or compromised home network that halt the STB-based DVR/PVR software (e.g. a kill command). Management commands may also be executed by an intruder able to perform management activities. This could also be accomplished by malware.

Overwriting of STB-based DVR/PVR software or data via a buffer overflow could cause the DVR/PVR software to hang, crash or terminate prematurely.

Forged trick play commands (e.g. STOP commands) received over the middleware VLAN or compromised home network destined for the STB-based DVR/PVR software could cause the DVR/PVR software prematurely to terminate recording/playback.

### 5.4.2.5 STB Credentials

There is a threat of unauthorized management activity resulting in the deletion of STB credentials. Unauthorized access to the administrative console can be used to send commands over the management VLAN or compromised home network that delete the STB credentials.

There is a threat of unauthorized management activity resulting in tampering with (or modification of) STB credentials. Unauthorized access to the administrative console can be used to send commands over the management VLAN or compromised home network that modify the STB credentials. This could also be the result of an intruder able to perform management activities or the result of software downloads from illegitimate sources.

Instructions and material necessary for cloning cable modems and cable STBs are readily available over the Internet. It is highly probable that this capability will also emerge for IPTV STBs.

Unauthorized duplication of STB credentials could also be the result of an intruder who is able to perform management activities. An intrusion may be the result of an external or internal attack.

Tampering with STB credentials could result in modification of their expiry to indicate they have not expired (e.g. in an unused STB), resulting in theft of IPTV service.

Publicly available tools can be used to discover the existence of STBs and potentially their vendor and version numbers, which can be used to search for known vulnerabilities that can be exploited to provide access to the STB, thereby allowing unauthorized browsing.

Tampering with STB credentials could result in modification of their expiry to indicate they have expired, resulting in interruption of the IPTV service.

Deletion of the STB credentials prevents the STB from authenticating with the IPTV service, resulting in interruption of the IPTV service.

### 5.4.2.6 Digital Certificate (Software Provider)

There is a threat of unauthorized management activity resulting in the deletion of the software provider digital certificate. Unauthorized access to the administrative console can be used to send commands over the management VLAN or compromised home network that delete the software provider digital certificate.

There is a threat of unauthorized management activity resulting in tampering with (or modification of) the software provider digital certificate. Unauthorized access to the administrative console can be used to send commands over the management VLAN or compromised home network that modify the software provider digital certificate (e.g. shorten its duration). This could also be the result of an intruder able to perform management activities or the result of software downloads from illegitimate sources. An intrusion may be the result of an external or internal attack. An internal attack would most likely be accomplished by a dishonest administrator. An external attack would first require the STB to be broken into.

A forged digital certificate (e.g. a self-signed digital certificate) could be sent by a rogue software provider in conjunction with illegitimate software in order to masquerade as a legitimate software provider.

A forged certificate revocation list (CRL) could be sent to the STB, revoking the digital certificate of a legitimate software provider. This could result in an interruption of software upgrades, patches, etc., from the legitimate software provider.

Deleting or tampering with the software provider digital certificate would prevent the STB from authenticating a legitimate software provider, thereby interrupting software upgrades, patches, etc.

### 5.4.2.7 STB Digital Certificate

There is a threat of unauthorized management activity resulting in the deletion of the STB digital certificate. Unauthorized access to the administrative console can be used to send commands over the management VLAN or compromised home network that delete the STB digital certificate. This could also be performed by an intruder able to perform management activities. It could also be accomplished by malware.

There is a threat of unauthorized management activity resulting in tampering with (or modification of) the STB digital certificate. Unauthorized access to the administrative console can be used to send commands over the management VLAN or compromised home network that modify the STB digital certificate. This could also be the result of an intruder able to perform management activities or the result of software downloads from illegitimate sources.

Instructions and material necessary for cloning cable modems and cable STBs are readily available over the Internet. It is highly probable that this capability will also emerge for IPTV STBs.

Unauthorized duplication of the STB digital certificate could also be the result of an intruder able to perform management activities.

Publicly available tools can be used to discover the existence of STBs and potentially their vendor and version numbers, which can be used to search for known vulnerabilities that can be exploited to provide access to the STB, thereby allowing unauthorized browsing.

A forged certificate revocation list (CRL) could be sent to the IPTV service provider, revoking the digital certificate of a legitimate STB. This could result in an interruption of IPTV service to a legitimate subscriber.

Deleting or tampering with the STB digital certificate would prevent the IPTV service provider from authenticating a legitimate STB, thereby interrupting IPTV service.

### 5.4.2.8 Public Keys (Used for Digital Certificates)

There is a threat of unauthorized management activity resulting in the deletion of stored public keys. Unauthorized access to the administrative console can be used to send commands over the management VLAN or compromised home network that delete the stored public keys from the STB.

There is a threat of unauthorized management activity resulting in tampering with (or modification of) or installation of stored public keys. Unauthorized access to the administrative console can be used to send commands over the management VLAN or compromised home network that insert a bogus public key or modify the stored public keys in the STB.

Deleting or tampering with the stored public keys would prevent the STB from authenticating the legitimacy of digital signatures and digital certificates, possibly resulting in interruption of IPTV service.

## 5.4.3 STB User Storage

### 5.4.3.1 Downloaded Content

There is a threat of unauthorized management activity resulting in the deletion of downloaded content. Unauthorized access to the administrative console can be used to send commands over the management VLAN or compromised home network that delete downloaded content.

There is a threat of unauthorized management activity resulting in tampering with (or modification of) downloaded content. Unauthorized access to the administrative console can be used to send commands over the management VLAN or compromised home network that modify downloaded content. This could also be the result of an intruder able to perform management activities or the result of software downloads from illegitimate sources.

If the downloaded content is stored in the clear, or if decryption keying material is accessible, a thief could make a bit-by-bit copy of the unencrypted downloaded content and distribute it to others, thereby depriving the content owner of revenue.

Tampering with the local copy of a subscriber's content rights stored in DRM storage could result in an invalid extension of the subscriber's ability to access the content, thereby depriving the content owner of revenue. Forged messages received by the DRM software could instruct the DRM software to extend the subscriber's content rights.

Publicly available tools can be used to discover the existence of STBs and potentially their vendor and version numbers, which can be used to search for known vulnerabilities that can be exploited to provide access to the STB, thereby allowing unauthorized browsing that would reveal the existence of downloaded content.

Tampering with the local copy of a subscriber's content rights could result in modifying their expiry to indicate they have expired, resulting in interruption of access to the downloaded content. Forged messages received by the DRM software could instruct the DRM software to expire the subscriber's content rights.

Deletion of the subscriber's content rights tricks the DRM software into believing that the subscriber has no rights for the downloaded content, resulting in an interruption of access to the downloaded content.

Unauthorized deletion or tampering with downloaded content prevents its playback.

### 5.4.3.2 User-created Content

User-created content is created by third-party applications that are optionally stored on the STB. Examples of this type of content would be organizing videos, photos, music, etc., into albums or themes.

There is a threat of unauthorized management activity resulting in the deletion of user-created content. Unauthorized access to the administrative console can be used to send commands over the management VLAN or compromised home network that delete user-created content.

There is a threat of unauthorized management activity resulting in tampering with (or modification of) user-created content. Unauthorized access to the administrative console can be used to send commands over the management VLAN or compromised home network that modify user-created content. This could also be the result of an intruder able to perform management activities or the result of software downloads from illegitimate sources. An intrusion may be the result of an external or internal attack. An internal attack would most likely be accomplished by a dishonest administrator. An external attack would first require the STB to be broken into.

Unauthorized deletion of or tampering with user-created content prevents its playback.

### 5.4.3.3 STB Smart Card

In this instance, the STB smart card is used to store authentication information.

### 5.4.3.4 STB Credentials

There is a threat of unauthorized management activity resulting in the deletion of STB credentials from the smart card. Unauthorized access to the administrative console can be used to send commands over the management VLAN or compromised home network that delete STB credentials from a smart card. It could also be performed by an intruder able to perform management activities. It could also be accomplished by malware.

There is a threat of unauthorized management activity resulting in tampering with (or modification of) STB credentials on the smart card. Unauthorized access to the administrative console can be used to send commands over the management VLAN or compromised home network that modify STB credentials on the smart card. This could also be the result of an intruder able to perform management activities or the result of software downloads from illegitimate sources.

Inserting a cloned STB smart card into a basic service STB would potentially allow it to receive enhanced IPTV services without being charged for them.

Instructions and material necessary for cloning cable modems and cable STBs are readily available over the Internet. It is highly probable that this capability will also emerge for IPTV STB smart cards.

In addition, it is possible that smart card copiers will exist on the black market.

Unauthorized duplication of the smart card could also be the result of an intruder able to perform management activities.

Publicly available tools can be used to discover the existence of STBs and potentially their vendor and version numbers, which can be used to search for known vulnerabilities that can be exploited to provide access to the STB, thereby allowing unauthorized browsing of STB smart cards.

Tampering with STB credentials could result in modification of their expiry to indicate they have expired, resulting in interruption of the IPTV service.

Deletion of the STB credentials prevents the STB from authenticating with the IPTV service, resulting in interruption of the IPTV service.

### 5.4.3.5 STB Digital Certificate

There is a threat of unauthorized management activity resulting in the deletion of the STB digital certificate from the smart card. Unauthorized access to the administrative console can be used to send commands over the management VLAN or compromised home network that delete the STB digital certificate from the smart card. It may also be performed by an intruder able to perform management activities.

There is a threat of unauthorized management activity resulting in tampering with (or modification of) the STB digital certificate on the smart card. Unauthorized access to the administrative console can be used to send commands over the management VLAN or compromised home network that modify the STB digital certificate on the smart card. This could also be the result of an intruder able to perform management activities or the result of software downloads from illegitimate sources.

Publicly available tools can be used to discover the existence of STBs and potentially their vendor and version numbers, which can be used to search for known vulnerabilities that can be exploited to provide access to the STB, thereby allowing unauthorized browsing of STB smart cards.

It is also possible that smart card copiers will exist on the black market.

A forged certificate revocation list (CRL) could be sent to the IPTV service provider, revoking the digital certificate of a legitimate STB. This could result in an interruption of IPTV service to a legitimate subscriber.

Deleting or tampering with the STB digital certificate would prevent the IPTV service provider from authenticating a legitimate STB, thereby interrupting IPTV service.

### 5.4.3.6 STB High-definition Output Interface

The underlying issue is that it is possible to connect a recording device to the high-definition output interface, which would allow making a bootleg copy of the high-definition content available on the Internet.

#### 5.4.3.7 DVI

Connecting a DVI recording device to the high-definition output interface would allow making a bootleg digital copy of the high-definition content available on the Internet.

The same is applicable to PCI, IEEE 1394 Serial Bus, DLNA UPnP and HDMI.

### 5.4.4 Residential Gateway

Instructions and material necessary for the creation and distribution of malware and rootkits are readily available over the Internet.

Once an attacker has broken into the residential gateway, he or she can perform arbitrary management commands and the integrity of the system can no longer be guaranteed.

Forged SNMP and CWMP commands (containing RPCs embedded in forged SOAP messages) can arrive via an insecure home network or via the management VLAN.

Publicly available tools can be used to discover the existence of a residential gateway and potentially its vendor and version number, which can be used to search for known vulnerabilities that can be exploited to provide access to the residential gateway, thereby allowing unauthorized browsing.

DOS attacks (e.g. malformed messages, traffic flood, buffer overflows) against the residential gateway can arrive via an insecure home network or any of the VLANs on the WAN interface.

Once an attacker has broken into the residential gateway, he or she can perform arbitrary management commands including shutting down the system.

### 5.4.5 DSLAM

The current scope consists of vulnerabilities unique to IPTV contained within the home network environment and the IPTV core network demarcation on the DSLAM.

#### 5.4.5.1 Audience Metering Information

There is a threat of unauthorized management activity resulting in the deletion of audience metering information. Unauthorized access to the administrative console can be used to send commands over the management VLAN or compromised home network that delete the audience metering information. Unauthorized deletion of audience metering information could also occur as a result of an internal or external attack or be caused by malware.

There is a threat of unauthorized management activity resulting in the modification of audience metering information. Unauthorized access to the administrative console can be used to send commands over the management VLAN or compromised home network that modify the audience metering information. Unauthorized tampering with audience metering information could also occur as a result of an internal or external attack or be caused by malware.

Unauthorized management activity resulting in the duplication of audience metering information. Unauthorized access to the administrative console can be used to send commands over the management VLAN or compromised home network that duplicate the audience metering information.

Unauthorized duplication of audience metering information could also occur as a result of an internal or external attack or be caused by malware.

Publicly available tools can be used to discover the existence of a DSLAM and potentially its vendor and version number, which can be used to search for known vulnerabilities that can be exploited to provide access to the DSLAM, thereby allowing unauthorized browsing of audience metering information.

### 5.4.5.2 Fraud Control Information

There is a threat of unauthorized management activity resulting in the deletion of fraud control information. Unauthorized access to the administrative console can be used to send commands over the management VLAN or compromised home network that delete the fraud control information. Unauthorized deletion of fraud control information could also occur as a result of an internal or external attack. For an external attack, a break-in of the DSLAM would be required first, and then the attacker would delete the fraud control information.

*Compromise of Platform Integrity*

There is a threat of unauthorized management activity resulting in the modification of fraud control information. Unauthorized access to the administrative console can be used to send commands over the management VLAN or compromised home network that modify the fraud control information. Unauthorized tampering with fraud control information could also occur as a result of an internal or external attack. For an external attack, a break-in of the DSLAM would be required first, and then the attacker would tamper with the fraud control information. For an internal attack, it is most likely that a dishonest administrator would tamper with the fraud control information. One possible motive for tampering with fraud control information would be to cover up the perpetration of a fraud.

The ultimate motive for deletion of or tampering with fraud control information would be to prevent detection of fraud in order successfully to steal IPTV service.

Publicly available tools can be used to discover the existence of a DSLAM and potentially its vendor and version number, which can be used to search for known vulnerabilities that can be exploited to provide access to the DSLAM, thereby allowing unauthorized browsing of fraud control information.

### 5.4.5.3 IP Filters

There is a threat of unauthorized management activity resulting in the deletion of IP filters. Unauthorized access to the administrative console can be used to send commands over the management VLAN or compromised home network that delete the IP filters. Unauthorized deletion of IP filters via a management command could occur as a result of an internal or external attack. For an external attack, a break-in of the DSLAM would be required first, and then the attacker would delete the IP filters. For an internal attack, it is most likely that a dishonest administrator would delete the IP filters. One possible motive for deleting IP filters would be to allow access to a video stream without paying for it.

A forged multicast protocol (e.g. IGMP) message could be sent to the DSLAM, instructing it to delete IP filters for the purpose of allowing access to a video stream without paying for it. Forged multicast protocol messages can originate from the home network environment or via injection into the broadcast TV VLAN or VOD VLAN.

There is a threat of unauthorized management activity resulting in the modification of IP filters. Unauthorized access to the administrative console can be used to send commands over the management VLAN or compromised home network that modify the IP filters. Unauthorized modification of IP filters via a management command could also occur as a result of an internal or external attack. For an external attack, a break-in of the DSLAM would be required first, and then the attacker would modify the IP filters. For an internal attack, it is most likely that a dishonest administrator would modify the IP filters. One possible motive for modifying IP filters would be to allow access to a video stream without paying for it. Conversely, modification of IP filters could also deny legitimate access to a video stream.

Publicly available tools can be used to discover the existence of a DSLAM and potentially its vendor and version number, which can be used to search for known vulnerabilities that can be exploited to provide access to the DSLAM, thereby allowing unauthorized browsing of IP filters.

IP filters could be modified to prevent legitimate access to a video stream.

## 5.4.6 Broadcast/Multicast TV VLAN Service

### 5.4.6.1 Decryption Keys

There is a threat of unauthorized management activity resulting in the deletion of media decryption keys. Unauthorized access to the administrative console can be used to send commands over the management VLAN or compromised home network that delete the media decryption key. Unauthorized deletion of the media decryption key could also occur as a result of an internal or external attack or be caused by malware.

There is a threat of unauthorized management activity resulting in the modification of media decryption keys. Unauthorized access to the administrative console can be used to send commands over the management VLAN or compromised home network that modify the media decryption key. Unauthorized modification of the media decryption key could occur as a result of an internal or external attack or be caused by malware.

If decryption keying material is accessible, a thief could make a copy of the subscriber's decryption key and use it to obtain content without paying for it.

Intruder activity of guessing and then obtaining vulnerable media decryption keys allows access to content without paying for it.

If decryption keying material is accessible, a subscriber could modify the key lifetime to obtain additional content, for example in order to view the content for a longer time period, etc., without paying for it.

Publicly available tools can be used to discover the existence of an STB and potentially its vendor and version number, which can be used to search for known vulnerabilities that can be exploited to provide access to the STB, thereby allowing unauthorized browsing that would reveal the existence of decryption keying information.

Decryption keying information could be captured in transit by means of a publicly available packet sniffer (e.g. ethereal) or MITM. These attacks would require access to the home network environment or the middleware VLAN between the DSLAM and the residential gateway.

If decryption keying material is accessible, an attacker could access the decryption keying material and insert an invalid key lifetime, resulting in interruption of access to valid content.

If a decryption key is deleted or modified, the DRM service will not allow access to authorized content.

### 5.4.6.2 CWMP

CWMP is SOAP-based protocol that uses embedded RPCs to manage CPE. The security vulnerabilities of CWMP have not been fully explored. However, SOAP is a text-based protocol.

Forged CWMP messages or unauthorized modification of CWMP messages in transit would most likely result in unauthorized management activities on the STB or residential gateway. This could be accomplished by injecting forged CWMP packets into the management connection of the STB or residential gateway or by an MITM attack. These attacks would originate from the home network environment or the management VLAN between the DSLAM and the residential gateway.

Forged or modified CWMP messages could cause the residential gateway to redirect content to an unauthorized subscriber, resulting in theft of content.

CWMP messages could be captured in transit by means of a publicly available packet sniffer (e.g. ethereal) or MITM. These attacks would require access to the home network environment or the management VLAN between the DSLAM and the residential gateway. A potential motive for capturing CWMP packets could be obtaining management authentication information contained in them.

Forged or modified CWMP messages could contain instructions that cause the STB or residential gateway to halt or reboot, resulting in IPTV service interruption.

A packet flood of illegitimate CWMP messages to the STB or residential gateway would cause the STB or residential gateway to stop responding. In addition, a traffic flood could consume all of the bandwidth on the STB or residential gateway management connection, preventing legitimate CWMP traffic from getting through. These attacks would originate from the home network environment or the management VLAN between the DSLAM and the residential gateway.

### 5.4.6.3 NTP/SNTP

There is a threat of injection of forged NTP/SNTP packets into the management connection of STBs, residential gateways or DSLAM or modification of NTP/SNTP packets in transit by an MITM attack. These attacks would originate from the home network environment or the management VLAN between the DSLAM and the residential gateway.

Pay-per-view multicasts, billing information, accountability information, etc., require tight network time synchronization. Forged or modified NTP/SNTP messages could cause the network time between devices to become unsynchronized, thereby interrupting IPTV service, for example by preventing a paid subscriber from joining a PPV multicast.

A packet flood of illegitimate NTP/SNTP packets to the STB or residential gateway could cause the STB or residential gateway to stop responding. In addition, a traffic flood could consume all of the bandwidth on the STB or residential gateway management connection, preventing legitimate CWMP traffic from getting through.

## 5.4.7 Broadcast/Multicast TV Application

### 5.4.7.1 MPEG-2 and MPEG-4 Video Stream

Malformed MPEG packets have the potential to result in a buffer overflow or cause the system to crash if improperly implemented. A specially crafted MPEG-2 stream can cause a buffer overflow in vulnerable software that allows remote attackers to execute arbitrary code. Multiple multimedia players (e.g. Quicktime, xine) are susceptible to these types of attack.

Forged MPEG packets or unauthorized modification of MPEG packets in transit would most likely result in the loss of fidelity of the MPEG stream. This could be accomplished by injecting noise into the MPEG stream or by an MITM attack.

An attacker could send a flood of MPEG packets to the STB in order to cause the STB to crash. Owing to the limited bandwidth of the STB interfaces, this could be a DOS (as opposed to a D-DOS). In addition, a traffic flood on the broadcast TV connection could consume all of its bandwidth and prevent legitimate traffic from getting through.

### 5.4.7.2 DSM-CC

DSM-CC is used to provide VCR-like controls for MPEG-2 flows. The security vulnerabilities of these types of protocol have not been fully explored.

Malformed DSM-CC packets have the potential to result in a buffer overflow or cause the video server to crash if improperly implemented, potentially interrupting the VOD service. The direction of the attack would be towards the video server and away from the STB. For the current scope, this attack would originate in either the home network environment or the VOD VLAN between the DSLAM and residential gateway.

In addition, modified/forged DSM-CC packets could result in bogus trick play commands being sent to the video server on behalf of the subscriber (e.g. bogus STOP commands), thus interrupting the VOD service.

An attacker could send a flood of DSM-CC packets to the video server in order to cause the video server to crash. This attack could originate from within an insecure home network environment (e.g. WLAN) or from injection into the VOD VLAN.

## 5.4.8 Middleware Application

### 5.4.8.1 EPG

Unauthorized deletion of the EPG from an STB would be accomplished by unauthorized management activity or malware. Unauthorized access to the administrative console can be used to send commands over the management VLAN or compromised home network that delete the EPG from the STB.

A potential motive for modifying or forging an EPG would be phishing attacks, where the subscriber is redirected to a bogus video server and prompted to enter personal information for the purpose of identity theft.

Unauthorized modification of the EPG in an STB would be accomplished by unauthorized management activity or malware. Unauthorized access to the administrative console can be used to send commands over the management VLAN or compromised home network that modify the EPG in the STB.

Forging the EPG or unauthorized modification of the EPG in transit could be accomplished by injecting packets into the connection between the STB and middleware server or by an MITM attack. These attacks would originate either from the home network environment or from the middleware VLAN between the DSLAM and the residential gateway.

There are a number of ways that an illegitimate service provider may pose as the authoritative source for the EPG, such as an MITM attack, DNS cache poisoning, etc.

### 5.4.8.2 Menus

Unauthorized deletion of a menu from an STB would be accomplished by unauthorized management activity or malware. Unauthorized access to the administrative console can be used to send commands over the management VLAN or compromised home network that delete the menu from the STB.

A potential motive for modifying or forging a menu would be phishing attacks, where the subscriber is redirected to a bogus video server and prompted to enter personal information for the purpose of identity theft.

Unauthorized modification of a menu in an STB would be accomplished by unauthorized management activity or malware. Unauthorized access to the administrative console can be used to send commands over the management VLAN or compromised home network that modify the menu in the STB.

### 5.4.8.3 Subscriber Credentials

Unauthorized deletion of subscriber credentials from an STB would be accomplished by unauthorized management activity or malware. Unauthorized access to the administrative console can be used to send commands over the management VLAN or compromised home network that delete the subscriber credentials from the STB. This management activity could also be the result of an external or internal attack.

Unauthorized insertion or modification of subscriber credentials in an STB would be accomplished by unauthorized management activity, malware or rootkit. Unauthorized access to the administrative console can be used to send commands over the management VLAN or compromised home network that insert or modify the subscriber credentials in the STB.

Unauthorized insertion or modification of subscriber credentials in transit could be accomplished by injecting packets into the connection between the STB and middleware server or by an MITM attack. These attacks would originate either from the home network environment or from the middleware VLAN between the DSLAM and the residential gateway.

A possible motive for the unauthorized insertion or modification of subscriber credentials would be the forgery of another subscriber's credentials in order to steal service.

Unauthorized copying of subscriber credentials from an STB would be accomplished by unauthorized management activity or malware. Unauthorized access to the administrative console can be used to send commands over the management VLAN or compromised home network that copy the subscriber credentials in the STB.

Unauthorized capture of subscriber credentials in transit could be accomplished by placing a packet sniffer device or using packet sniffing software in the connection between the STB and middleware server or by an MITM attack.

There are a number of ways that an illegitimate service provider may pose as a legitimate service provider, such as an MITM attack, DNS cache poisoning, etc.

The ultimate motive for all of these threats is identity theft.

Placing a packet sniffer device or using packet sniffing software in the connection between the STB and middleware server could result in disclosure of subscriber credentials. These attacks would originate either from the home network environment or from the middleware VLAN between the DSLAM and the residential gateway.

Publicly available tools can be used to discover the existence of STBs and potentially their vendor and version numbers, which can be used to search for known vulnerabilities that can be exploited to provide access to the STB, thereby allowing unauthorized browsing.

The unauthorized deletion or modification of subscriber credentials could result in the middleware server being unable to authenticate the subscriber, resulting in interruption of IPTV service.

### 5.4.8.4 Purchasing Information

Unauthorized deletion of purchasing information from an STB would be accomplished by unauthorized management activity or malware. Unauthorized access to the administrative console can be used to send commands over the management VLAN or compromised home network that delete the purchasing information from the STB.

Unauthorized insertion or modification of purchasing information in an STB would be accomplished by unauthorized management activity, malware or rootkit. Unauthorized access to the administrative console can be used to send commands over the management VLAN or compromised home network that insert or modify purchasing information in the STB.

Unauthorized insertion or modification of purchasing information in transit could be accomplished by injecting packets into the connection between the STB and middleware server or by an MITM attack. These attacks would originate either from the home network environment or from the middleware VLAN between the DSLAM and the residential gateway.

A possible motive for the unauthorized insertion or modification of purchasing information would be the forgery of another subscriber's purchasing information in order to avoid paying for service or content.

Unauthorized copying of purchasing information from an STB would be accomplished by unauthorized management activity or malware. Unauthorized access to the administrative console can be used to send commands over the management VLAN or compromised home network that copy purchasing information in the STB.

Unauthorized capture of purchasing information in transit could be accomplished by placing a packet sniffer device or using packet sniffing software in the connection between the STB and middleware server or by an MITM attack. These attacks would originate either from the home network environment or from the middleware VLAN between the DSLAM and the residential gateway.

There are a number of ways that an illegitimate service provider may pose as a legitimate service, such as an MITM attack, DNS cache poisoning, etc.

The ultimate motive for all of these threats is identity theft.

Placing a packet sniffer device or using packet sniffing software in the connection between the STB and middleware server could result in disclosure of subscriber credentials. These attacks would originate either from the home network environment or from the middleware VLAN between the DSLAM and the residential gateway.

Publicly available tools can be used to discover the existence of STBs and potentially their vendor and version numbers, which can be used to search for known vulnerabilities that can be exploited to provide access to the STB, thereby allowing unauthorized browsing.

### 5.4.8.5 Digital Certificates (Content Provider)

There is a threat of unauthorized management activity resulting in the deletion of the content provider digital certificate. Unauthorized access to the administrative console can be used to send commands over the management VLAN or compromised home network that delete the content provider digital certificate from the STB.

There is a threat of unauthorized management activity resulting in tampering with (or modification of) the content provider digital certificate. Unauthorized access to the administrative console can be used to send commands over the management VLAN or compromised home network that modify the content provider digital certificate in the STB.

A forged digital certificate (e.g. a self-signed digital certificate) could be sent by a rogue content provider in conjunction with illegitimate content in order to masquerade as a legitimate content provider.

### 5.4.8.6 Parental Controls

Unauthorized deletion of parental controls from an STB would be accomplished by unauthorized management activity or malware. Unauthorized access to the administrative console can be used to send commands over the management VLAN or compromised home network that delete the parental controls in the STB.

In addition, an attacker masquerading as a parent could delete the parental controls from the STB, thereby allowing access to inappropriate content.

Unauthorized insertion or modification of parental controls in an STB would be accomplished by unauthorized management activity, malware or rootkit. Unauthorized access to the administrative console can be used to send commands over the management VLAN or compromised home network that insert or modify parental controls in the STB.

Unauthorized insertion or modification of parental controls in transit could be accomplished by injecting packets into the connection between the STB and middleware server or by an MITM attack. These attacks would originate either from the home network environment or from the middleware VLAN between the DSLAM and the residential gateway.

Parental controls may be a target on the basis of the fact that many people use the same password/PIN for different things.

Unauthorized copying of parental controls from an STB would be accomplished by unauthorized management activity or malware. Unauthorized access to the administrative console can be used to send commands over the management VLAN or compromised home network that copy the parental controls in the STB.

Unauthorized capture of parental controls in transit could be accomplished by placing a packet sniffer device or using packet sniffing software in the connection between the STB and middleware server or by an MITM attack. These attacks would originate either from the home network environment or from the middleware VLAN between the DSLAM and the residential gateway.

There are a number of ways that an illegitimate service provider may pose as a legitimate service, such as an MITM attack, DNS cache poisoning, etc.

Placing a packet sniffer device or using packet sniffing software in the connection between the STB and middleware server could result in disclosure of parental controls. These attacks would originate either from the home network environment or from the middleware VLAN between the DSLAM and the residential gateway.

Publicly available tools can be used to discover the existence of STBs and potentially their vendor and version numbers, which can be used to search for known vulnerabilities that can be exploited to provide access to the STB, thereby allowing unauthorized browsing.

A possible motive for the unauthorized insertion of parental controls would be to deny access to authorized content. This would be a nuisance attack.

### 5.4.8.7 PVR/DVR Application

Unauthorized deletion of recorded content stored in an STB would be accomplished by unauthorized management activity or malware. Unauthorized access to the administrative console can be used to send commands over the management VLAN or compromised home network that delete recorded content stored in the STB.

Unauthorized insertion or modification of recorded content stored in an STB would be accomplished by unauthorized management activity, malware or rootkit. Unauthorized access to the administrative console can be used to send commands over the management VLAN or compromised home network that insert or modify recorded content stored in the STB.

If the recorded content is stored in the clear, or if decryption keying material is accessible, a thief could make a bit-by-bit copy of the unencrypted recorded content and distribute it to others, thereby depriving the content owner of revenue.

Publicly available tools can be used to discover the existence of STBs and potentially their vendor and version numbers, which can be used to search for known vulnerabilities that can be exploited to provide access to the STB, thereby allowing unauthorized browsing that would reveal the existence of recorded content stored in the STB.

Unauthorized deletion or modification of recorded content stored in the STB prevents its playback.

### 5.4.8.8 User-sourced Content

User-sourced content would be created, for example, by connecting a video feed to the PVR/DVR.

Unauthorized deletion of user-sourced content stored in an STB would be accomplished by unauthorized management activity or malware. Unauthorized access to the administrative console can be used to send commands over the management VLAN or compromised home network that delete the user-sourced content in the STB.

Unauthorized insertion or modification of user-sourced content stored in an STB would be accomplished by unauthorized management activity, malware or rootkit. Unauthorized access to the administrative console can be used to send commands over the management

VLAN or compromised home network that insert or modify user-sourced content stored in the STB.

Unauthorized copying of user-sourced content stored in an STB would be accomplished by unauthorized management activity or malware. Unauthorized access to the administrative console can be used to send commands over the management VLAN or compromised home network that copy user-sourced content stored in the STB.

Unauthorized capture of user-sourced content in transit to a network-based PVR/DVR could be accomplished by placing a packet sniffer device or using packet sniffing software in the connection between the STB and network-based PVR/DVR or by an MITM attack. These attacks would originate either from the home network environment or from between the DSLAM and the residential gateway on the VLAN used for the PVR/DVR (possibly the VOD VLAN).

Placing a packet sniffer device or using packet sniffing software in the connection between the STB and PVR/DVR could result in disclosure of user-sourced content. These attacks would originate either from the home network environment or from between the DSLAM and the residential gateway on the VLAN used for PVR/DVR (possibly the VOD VLAN).

## 5.4.9 Application Management

### 5.4.9.1 IPTV Usage Information

IPTV usage information would be used to generate bills for usage-based services.

Unauthorized deletion of IPTV usage information stored in an STB or DSLAM would be accomplished by unauthorized management activity or malware. Unauthorized access to the administrative console can be used to send commands over the management VLAN or compromised home network that delete IPTV usage information stored in the STB or DSLAM.

Unauthorized modification of IPTV usage information stored in an STB or DSLAM would be accomplished by unauthorized management activity, malware or rootkit. Unauthorized access to the administrative console can be used to send commands over the management VLAN or compromised home network that modify IPTV usage information stored in the STB or DSLAM.

Unauthorized deletion or modification of subscriber IPTV usage information, whether in the STB, DSLAM or in transit between the STB and DSLAM, could result in the subscriber not being charged for IPTV services.

Unauthorized duplication of IPTV usage information stored in an STB or DSLAM would be accomplished by unauthorized management activity or malware. Unauthorized access to the administrative console can be used to send commands over the management VLAN or compromised home network that duplicate IPTV usage information stored in the STB or DSLAM.

Unauthorized interception of IPTV usage information in transit from the STB to DSLAM could be accomplished by placing a packet sniffer device or using packet sniffing software in the connection between the STB and DSLAM or by an MITM attack. These attacks would originate either from the home network environment or from the management VLAN between the DSLAM and the residential gateway.

One possible motive for duplicating subscriber IPTV usage information would be to sell this information.

Publicly available tools can be used to discover the existence of STBs and DSLAMs and potentially their vendor and version numbers, which can be used to search for known vulnerabilities that can be exploited to provide access to the STB or DSLAM, thereby allowing unauthorized browsing that would reveal the existence of subscriber IPTV usage information stored in the STB or DSLAM.

Placing a packet sniffer device or using packet sniffing software in the connection between the STB and DSLAM could result in disclosure of subscriber IPTV usage information. These attacks would originate either from the home network environment or from the management VLAN between the DSLAM and the residential gateway.

### 5.4.9.2 IPTV Billing Information

IPTV billing information is created when a subscriber views on-demand content such as PPV or VOD.

Unauthorized deletion of IPTV billing information stored in an STB or DSLAM would be accomplished by unauthorized management activity or malware. Unauthorized access to the administrative console can be used to send commands over the management VLAN or compromised home network that delete IPTV billing information stored in the STB or DSLAM. Unauthorized deletion of IPTV billing information stored in the STB or DSLAM could also occur as a result of an internal or external attack.

Unauthorized modification of IPTV billing information stored in an STB or DSLAM would be accomplished by unauthorized management activity, malware or rootkit. Unauthorized access to the administrative console can be used to send commands over the management VLAN or compromised home network that modify IPTV billing information stored in the STB or DSLAM. Unauthorized modification of IPTV billing information stored in the STB or DSLAM could also occur as a result of an internal or external attack.

Unauthorized deletion or modification of subscriber IPTV billing information, whether in the STB, DSLAM or in transit between the STB and DSLAM, could result in the subscriber not being charged for IPTV services.

Unauthorized duplication of IPTV usage information stored in an STB or DSLAM would be accomplished by unauthorized management activity or malware. Unauthorized access to the administrative console can be used to send commands over the management VLAN or compromised home network that duplicate IPTV billing information stored in the STB or DSLAM.

Unauthorized interception of IPTV billing information in transit from the STB to DSLAM could be accomplished by placing a packet sniffer device or using packet sniffing software in the connection between the STB and DSLAM or by an MITM attack. These attacks would originate either from the home network environment or from the management VLAN between the DSLAM and the residential gateway.

One possible motive for duplicating subscriber IPTV billing information would be to sell this information.

Placing a packet sniffer device or using packet sniffing software in the connection between the STB and DSLAM could result in disclosure of subscriber IPTV billing information. These attacks would originate either from the home network environment or from the management VLAN between the DSLAM and the residential gateway.

## 5.5  Conclusion

The IPTV environment is formed by a high number of components that present security vulnerabilities at different levels. Operating systems, applications and protocols could have security problems that may allow intruders to take control of the environment or cause operating failures.

It is paramount that security professionals undertake an end-to-end security review of their IPTV environment, including the design phase, deployment and operations. Security countermeasures would have a better result if based on specific reviews and analysis of the infrastructure.

A viable approach to analyzing the threats is to segment the environment into the head-end, transport and home-end areas; this way, different teams could take ownership of reviewing and implementing countermeasures.

# 6

# Countering the Threats

Covering all the security requirements of an IPTV environment brings a high level of complexity. Security professionals must take a risk-based approach and deploy countermeasures that would reduce the exposure on the most critical assets. There should be a balance between the expected costs of the countermeasures and the benefit in terms of risk reduction. The countermeasures in this chapter are aimed at the creation of a number of security layers that provide complementary protection. Most of these countermeasures are interlinked, and, when one is breached, other countermeasures will still be available to protect the IPTV environment.

This chapter is arranged in three main areas of the IPTV environment: head end, home end and aggregation/transport. It will provide detailed recommendations on the controls that should be deployed, as well as how these controls will impact upon the threat areas presented in the previous chapter. Some countermeasures span across different areas. For example, IGMPv3 would impact upon the three areas, but for simplicity it will be considered only in relation to the aggregation and transport network.

As with the threats, this book will not define specific brands or products that should be used. Instead, the chapter will cover general solutions or controls that are required, allowing security professionals to select the available solution that best fits their needs. Products and brands tend to change over time and will lose validity. Security professionals can look for functions instead and apply this template to products they plan to source.

## 6.1 Securing the Basis

### 6.1.1 Hardening Operating Systems

Most elements within the IPTV environment will be supported by commercial operating systems. This will be a common situation within the head end, transport network and even within the home end. In some cases, set top boxes will be running on proprietary operating systems and security information will be limited. However, the market tendency is to have open-source or commercial off-the-shelf products that can be used, facilitating increased flexibility and compatibility with hardware platforms and DRM/IPTV clients.

For all the components running on known operating systems, security professionals must ensure that the systems have been properly configured and accounted for and have been patched for known security problems. These three aspects of security tend to fail within large deployments and specifically when applications or components are changing frequently. An IPTV environment may have hundreds of servers and hundreds of thousands of set top boxes. Clearly, manual configuration and patching is not a task considered economically viable.

The IPTV platform brings a new set of applications and many additional servers that must be protected, and this could take its toll on the resources available for operations and maintenance of the platform. Having IT support personnel manually configuring all systems may create a significant overhead and also a long response time to known security problems. One option available to reduce the amount of time and resources required for hardening the operating systems is to use automated tools to validate the security posture of the platform and also to deploy security configurations.

There are existing security vulnerabilities that would allow intruders easily to take control of the middleware server via web service exploits. Intruders would then be able to use the middleware server as a stepping stone to attack other internal systems, which will bypass some of the security mechanisms deployed internally.

Regardless of the number of security layers implemented in an IPTV environment, and irrespective of the investment in security mechanisms, security will be broken if operating systems and applications are not patched and properly configured.

The Department of Homeland Security in the USA has funded the development of a number of projects supporting the technology required to automate the process of securing a computing platform [1]. The Mitre Corporation is developing a number of initiatives relevant to this goal, including:

- common vulnerabilities and exposures (CVE®);
- common platform enumeration (CPE™);
- common configuration enumeration (CCE™).

CVE® provides frequent updates describing vulnerabilities found on operating systems, and some major services and applications. This information includes a common numbering and classification for vulnerabilities, allowing for identification of security issues using a common language. CVE® is also improved by the work from the national vulnerability database (NVD), which is a product of the US National Institute of Standards and Technology (NIST) Computer Security Division and includes threat classification for known vulnerabilities. Using CVE®, security professionals will be able to determine which patches and modifications are required by a particular system. This information is extracted on the basis of the current configuration. The NVD entries are provided as XML feeds and are also available for online browsing. CVE® provides a consistent naming of security vulnerabilities, which is critical if security professionals want to have a unified management of security across the IPTV environment. Using CVE®, different software tools can be employed to assess the security of the platform, and results will be equivalent in terms of naming. More flexibility is added to the process, as different tools can be used and results can be compared on a like-for-like basis.

CPE™ allows automatic inventories of computer networks by providing a common naming structure and some elements for recognition of operating systems. This can be used to create an XML representation of the inventory of systems within the network. With CPE™, security professionals can create a common inventory representing all elements in the IPTV environment and match that inventory with the existing vulnerabilities known on those systems. As new platforms are added, the inventory can be adapted to reflect changes in technology. This tool can be used as the basis of the inventory engine, allowing security professionals to have a clear view of the number and type of elements available within the IPTV environment. By having an updated inventory, the process of ensuring security for the environment is simplified. During the inventory audits, security professionals will be able to detect unauthorized elements within the infrastructure. In some cases, third-party vendors, system administrators or network administrators add systems or change their configuration. Systems are reinstalled without all the appropriate software patches, or servers are replaced without the required security configuration. Using CPE™, security professionals will be able to maintain up-to-date lists of systems and their current security level.

CCE™ provides an XML representation of the configuration of operating systems and major applications. This information includes the known security issues within operating systems as well as recommended security parameters that must be configured. In general, the CCE™ provides a common naming structure for software configuration issues. For example, it provides a common language to refer to password configuration parameters across all platforms. With CCE™, security professionals are able to document the specific security configuration required for each platform or for each platform profile. This information, added to the list of CVE® entries applicable on a platform, would present the complete picture of known security problems on a system. Security configuration problems facilitate unauthorized access and disruption of systems. By using CCE™, security professionals can define a detailed configuration model (or checklist) describing all the parameters that must be set on the available systems. Even within systems providing different services there are similar functions that require configuration. For example, password length requirements apply to operating systems and applications alike. CCE™ can be used to translate compliance requirements against the specific configuration parameters on each operating system. Security professionals can take an international standard such as ISO/IEC 17799 or an industry standard such as PCI DSS and translate most of the requirements into CCE™ language. Once the standard is in CCE™ form for a particular platform, it will act as a template that can be reused for all similar platforms (one template for each operating system and application; patches may not require a new template).

Additionally, the Forum of Incident Response and Security Teams (FIRST^SM) is working on the common vulnerability scoring system (CVSS-SIG™) providing a measurement of the criticality of vulnerabilities and their expected impact on the infrastructure. CVSS-SIG™ can be used to estimate the level of risk that a particular vulnerability presents. This information can also be used to select which vulnerabilities would be covered first and also the type of countermeasures more likely to provide a higher return on investment.

The information from CVSS-SIG™ can be considered an initial input. It considers the criticality of the vulnerability from a general perspective, but it will not provide the specific threat levels for an IPTV environment. Security professionals must add information about the asset value and exposure level. When comparing two servers with the same applications, operating system and asset value, if one is hosted on a DMZ exposed to the Internet and the

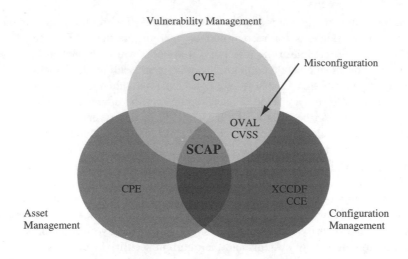

**Figure 6.1**  Integration of SCAP

other has only one physical connection linked with the video repository behind six layers of security, the two will suffer a different impact from the same vulnerability.

Figure 6.1 illustrates the common areas between CVE, CPE and CCE provided under SCAP.

All these projects are available on the Internet (http://nvd.nist.gov/scap.cfm) and can be integrated within a number of commercial off-the-shelf applications, using XML to coordinate information, allowing for automated configuration and audits. There are other corporate solutions that provide analysis of configuration and threats. Each IPTV service provider must select the one that better suits its needs and resources.

NIST has developed a number of checklists intended to provide security information for operating systems and widely used applications. Security checklists contain instructions for configuring or auditing a particular system. Checklists can be automated, and that automation supports the tasks required by security professionals operating in an environment requiring various efforts of compliance with laws, regulations and guidelines, as well as an accelerated rate of vulnerability discovery. Consistency is another benefit from using the checklists, as this ensures that, even when different individuals are involved in the configuration and auditing, the same result will be achieved.

Checklists are available for most major operating systems. The checklists can also be automated with scripts that are either documented on the checklist or provided by vendors. Security professionals can integrate these scripts and configure the platform to prevent specific attacks. Additionally, the scripts can be used while rebuilding a platform or after a programmed system refresh.

The security content automation program (SCAP) from NIST is aimed at supporting the development of automated checklists, compatible with the extensible configuration checklist description format (XCCDF) and/or the open vulnerability and assessment language (OVAL). XCCDF is used to define policies and configuration requirements, and OVAL is used for auditing the configuration of a particular platform.

Translating all this into IPTV and putting all elements together, security professionals will be able to use XML-based CPE™ automatically to gather an inventory of all elements within the IPTV environment. This inventory can then be checked against CVE® and the NVD to see the type of vulnerabilities that existing operating systems may have. Additionally, the list of vulnerabilities and information from the NVD can be augmented by the CVSS to determine the impact of the security vulnerabilities. OVAL can then be used to audit the systems and confirm security vulnerabilities. This can be repeated each time the platform is patched or reconfigured.

The technology is available for security professionals to create audit mechanisms that check if set top boxes have the appropriate security levels, and quarantine those that fail to meet the requirements. When a set top box is requesting a valid IP address through the DSLAM, the RADIUS server can verify through one of the local inventories if that particular set top box MAC or ID has been verified and complies with all the security policies. In some cases, the set top box can be added to a particular network segment with limited access until it has been audited and patched. Obviously, the security process cannot affect the customer experience, and any service disruptions must be avoided. Subscribers will not accept that their TV is not working because the IPTV service provider is patching the system.

Software vulnerabilities are not the only security problems that should be considered when hardening a platform. Misconfigurations may provide even easier access by intruders. The NIST configuration lists, translated using the XCCDF, can then be used to audit the configuration of the systems. Using OVAL as an auditing language and CCE™ as a common configuration language for the systems, security experts would be able to ascertain the level of compliance against expected configuration levels and issue detailed reports of configuration changes required.

Using all the elements together, based on an XML schema, would reduce the time and efforts required for maintaining the security of operating systems and major applications. Configuration changes and patches could be pushed automatically to all operating systems once they have been homologated for the IPTV environment.

Hardening would include activities such as (but not limited to) the following:

- patch implementation;
- removal of unnecessary applications and services;
- removal of unnecessary user accounts;
- password restrictions;
- log file and audit activation;
- implementation of access controls;
- white list of authorized applications;
- default file and folder permission.

For all servers, network elements and set top boxes, patch implementation is one of the most critical activities to maintain a secure environment. Security professionals must have a lab to test all recommended patches in a live environment and sign off the implementation of patches. Within set top boxes it is critical to ensure that changes will not affect the operation of the service. Different hardware models must be replicated in the lab environment.

Operating systems tend to include a large number of services that are not essential for the operation of the services. Set top boxes using open-source operating systems tend to have email services, web services and other TCP/IP-related services that are not required for the basic functions of the set top box. These additional services open the door to intruders and expose the system to unauthorized access by them. By removing unnecessary services there will be more resources available (in terms of memory and CPU) for other activities and fewer entry points for intruders, worms and viruses.

When installed by default, most operating systems will add unnecessary user accounts that could be used by intruders to have unauthorized access to the system. It is important that, during system installation, these accounts are removed.

. All operating systems and applications must be configured to require minimum levels of password complexity. This will ensure that passwords are not trivial and intruders cannot guess the passwords using default values or simple words. Remote access to set top boxes must require a valid password, even if the access can only be started from the home end. Network restrictions cannot replace strong password policies, as both must work together.

All activities within the IPTV environment must be properly recorded. This is done using the audit and log file functions from the operating systems and applications. For critical elements such as the head-end servers there should be a dedicated facility to store log files. There is available technology that can be used to send each log entry to a central repository, protecting audit information from unauthorized modification by intruders and ensuring that future investigations will have all the information required to reconstruct security incidents. These servers must be outside the control of the system administrators in charge of the operating systems and applications, as there could be a conflict of interest if the same individual were operating the systems and protecting the log files.

The operating systems must have access control restrictions blocking unauthorized access to all services and functions. No guest access or anonymous access should be allowed.

As all applications are known and approved by the IPTV service provider, white lists can be deployed within the server and set top box environment. Each application would have a specific fingerprint/checksum that could be verified by the operating system before allowing the application to be executed. This will block some viruses and rootkits, as they will not be authorized to be executed. Any unauthorized modification to parts of the operating system or applications may be recognized and blocked when compared against the white list.

Operating systems and applications create a number of local folders with default permissions. These tend to be inappropriate for secure operation of the applications and can be used by intruders to execute code or modify files. Proper permissions must be assigned to all folders and files, in accordance with the requirements for safe operation.

It is important to remember that all operating systems and major applications require hardening, regardless of the security mechanisms recommended in this chapter. If an intruder is able to exploit a security vulnerability or misconfiguration within one of the servers or applications, then the whole security model will be open to abuse. In most cases in this chapter, no additional references will be made to hardening-related activities in the recommended countermeasures.

## 6.1.2 Business Continuity

One of the key characteristics of TV services is that they are expected to be always on. Subscribers will not accept sporadic/irregular services from their IPTV service provider, and neither will they accept that network issues are responsible for a loss of service. From their point of view, both the IPTV service provider and the network are one and the same, and, more importantly, service is expected $24 \times 7$.

To ensure this level of service, particular attention should be directed towards defining a business continuity/disaster recovery strategy. This should include all elements, from the content acquisition to the home end, and must ensure that recovery of service is attained within acceptable timeframes. The scope of the plan must include an analysis of the most critical elements within the IPTV environment and also should create a layered approach that will allow security professionals to define how to protect the service.

VOD and broadcast TV rely on a common platform to deliver services. In some cases the only differences between the two services will be the VOD server. Security professionals must allocate resources to provide high availability and resilience within the IPTV platform. Servers must be deployed in high-availability mode when possible, and applications must be able to recover quickly from security incidents.

The transport network must be frequently evaluated to detect problems or malfunctions, allowing for timely reaction and modifications to either configuration or paths. Some elements may not be as critical as others. Criticality would be linked with the effect that a failure may cause. DSLAMs tend to cover between 5000 and 10 000 subscribers, but, with a subscriber base of 1 million, a DSLAM may not be considered a highly critical element.

All servers must be part of a permanent backup process, ensuring that the systems can be recovered to an operational stage on the basis of the stored information. Plans should be created to ensure that, in case of emergency, servers and applications can be recovered to an operational state. Frequent tests should be carried out to ensure that the plans are still viable and personnel involved in the recovery process are knowledgeable and ready.

Keys and critical components from the DRM service must be properly secured while in the backup tapes, using encryption to protect them from unauthorized access in case the tape is stolen or misplaced.

### 6.1.3 Intrusion Detection/Intrusion Prevention

All traffic within the IPTV environment must be monitored to detect known attacks and attempted intrusions. Worms, viruses and intruders have a number of alternative ways of entry, and security professionals must deploy specific elements to detect that behavior and take timely actions. Although firewalls and ACLs will be deployed across the IPTV environment, reducing the type of protocols and sessions allowed, it is important to maintain rules on the IDS/IPS that will detect sessions on ports and services that have been blocked by the firewall. This will act as a second mechanism to detect intrusions and also to find when one of the filters is failing, has been disabled or one of the administrators has made a mistake with the configuration of the system.

From the host-based IDS/IPS perspective there are different critical systems that should be protected:

- video repository;
- DRM server;
- middleware server;
- video streaming server;
- video on demand.

This list represents either the most exposed or the most valuable servers, and they must have an HIDS. The video repository holds the most valuable assets, and any intrusion may cause significant losses to the IPTV service providers. DRM servers hold the encryption keys to all content. Intruders will be interested in stealing the keys either to open encrypted content or to falsify communications to set top boxes. The middleware server is a central element in the operation of the platform and will be one of the first targets hit by attackers, as it allows communication from all set top boxes. The VOD and video streaming server will also accept communications from set top boxes, but only in limited ports. Intruders may plan to send denial of service or buffer overflows.

As secondary targets, with less exposure and with more difficult access by intruders, the following systems could also have an HIDS:

- MPEG encapsulator;
- transcoder;
- content management server;
- business servers.

From a network perspective, switches can be configured to mirror all traffic into a port that is connected to a network IDS/IPS. This will facilitate detection and containment of attacks.

At a minimum, the following segments must be included on the N-IDS/IPS scope:

- traffic from the home end to the head end (all VLANs);
- traffic on the middleware content management VLAN;
- traffic on the video repository content management VLAN;
- traffic on the DRM middleware VLAN.

Security professionals must decide which other VLANs are relevant owing to their particular environment and selected products.

Both network and host-based IDS/IPS systems must be properly configured to detect attacks targeted at the IPTV environment. Rules should be configured to detect specific behaviors that would affect the IPTV environment and services. One example could be to block administrative sessions/authentication over the VLAN providing communication between the set top box and the middleware server. Similar rules can be set up to block administrative/authentication traffic between the middleware and the content management server. The main purpose of these changes is to facilitate visibility over the behavior and trends within the head end and to allow security professionals to react in a timely fashion to attacks.

Additionally, security professionals must be aware of the effect that new rules will have on the flow of information. Some rules may block valid sessions, and this could eventually create an unintended service disruption. Tests should be undertaken to confirm that new rules are appropriate.

## 6.1.4 Network Firewalls

Network firewalls (or equivalent stateful inspection engines/application-aware engines) must be used to control the traffic within the head end. The traffic flow between the servers within a VLAN is known and documented, and hence a network firewall or equivalent mechanism can be deployed to ensure that only valid requests are transmitted. Some switches support ACLs and stateful inspection. These can be used for the purpose of enforcing the VLANs and also checking the traffic. If the switches are not a reliable mechanism, then dedicated equipment must be used for filtering and validating the traffic.

The firewalls and equivalent mechanisms will reduce the traffic between network elements and will facilitate the process of detecting unauthorized access. Initial configuration of the firewalls will require some support from the vendors providing IPTV components, as some applications may require more ports than originally documented.

Network firewalls tend to be used within the low-bandwidth connections, for example inside the head end or controlling the upstream traffic on the middleware VLAN. High-bandwidth traffic such as that from the video streaming server towards the set top boxes on the broadcast TV VLAN tends to be left without protection owing to the current inability of firewalls to withstand that level of traffic. In that particular case, switches and routers (and even the DSLAM) can be used to provide filtering functions.

## 6.1.5 Fraud Prevention

IPTV services must have fraud prevention capabilities. These can prevent abuse of the environment by subscribers or insider fraud. There are many different elements that could participate in fraud prevention and detection:

- The middleware can report activities from subscribers and detect when the same subscriber is requesting content using two separate IP addresses. This is a good example of set top box cloning, and it is relatively easy to detect by combining different antifraud mechanisms.
- The middleware server can detect when a set top box is requesting a very high number of VOD titles. This could represent unauthorized access to the set top box. A trigger can be established; for example, more than four titles in a day or more than eight titles in a week. If the set top box has been hijacked by intruders and they are downloading as many titles as possible, this mechanism will flag up suspicious behavior from a particular set top box.
- The RADIUS server can detect when the same subscriber is requesting access from two different IP addresses. This will work even if both are on separate DSLAMs. This mechanism can detect set top box cloning.
- The DSLAM can detect when the same subscriber is requesting access from two different physical lines (only if both are on the same DSLAM).
- Business servers and the RADIUS server can perform validation of users, establishing if there are any subscribers on the RADIUS system that have not been created on the business servers (billing, provisioning). This is a typical case where inside fraud is involved. In general, audits should be undertaken to confirm if all IPTV elements have the same number of subscribers and any anomaly should be investigated. Internal fraud can be detected by comparing records and finding which new accounts have been created.
- Business servers and the middleware server can perform a validation of users, establishing if there are any subscribers communicating with the middleware server who have not been created on the business server. This is a typical case where inside fraud is involved.

- DSLAM servers can report to the business servers which unicast titles have been received by subscribers. This information can be matched with the billing records to determine if there has been manipulation of the system.

## 6.1.6 DRM–CAS

As presented in Chapter 4, the IPTV service relies on DRM and CAS applications to protect the digital assets and implement a sustainable business model.

DRM applications must be evaluated to confirm that algorithms are secure and the implementation offers the appropriate safeguards.

## 6.2 Head End (IPTV Service Provider)

The central element of an IPTV infrastructure is the head end. This is formed by a number of elements that receive content, transform it and redistribute it to subscribers following the business model and subscriber packages available. The head end can be deployed using a central head end and regional head ends. This facilitates the broadcast of contents, as the regional head ends are closer to subscribers and latency is reduced.

### 6.2.1 Critical Elements of the Head End

*(i) Satellite Receivers*
Integrated receiver decoder (IRD).

*(ii) Video Repository*
This includes:

- video library;
- media library;
- library servers;
- storage area network;
- video-on-demand movie database;
- film server (video and audio files).

*(iii) Content Management System*
This includes:

- command center;
- asset management system;
- digital rights management.

*(iv) Master Video Streaming/Game Server*
This includes:

- propagation service;
- streaming service.

*(v) Ingest Gateway (Video Capture)*
This includes:

- recording system;
- recording manager;
- capture/distribution server.

*(vi) Video Cache Streaming Server*
This includes:

- caching server;
- media cluster.

*(vii) Middleware*
Middleware servers.

*(viii) Business-related Systems*
- accounting;
- provisioning;
- customer information.

The head end receives a series of data feeds in different formats and media including live feeds from local studios, premium contents from third parties, retransmissions, satellite contents and local video information. Owing to the variety of sources, some of this content is in analog mode and as such could not be transmitted to the IP network. Contents can be received in DVDs, tapes or video feeds via satellite or over-the-air communications. Contents are then encoded and encapsulated to allow TCP/IP-based transmission. Additional elements involved are the DRM application and content management systems. All communications with the subscribers are coordinated by the middleware server receiving requests from the different set top boxes.

The different components of the head end are illustrated in Figure 6.2. Security filters must be established to ensure protections over the communications within the head end.

## 6.2.2 Content Input

Video assets can be found in different media and formats within the head-end environment, and each type will have different specific threats and countermeasures related. Some of the countermeasures required to reduce the chance of unauthorized access to assets are as follows.

### 6.2.2.1 Satellite Feed

Content aggregators and some content owners distribute contents via satellite links. IPTV service providers use satellite receivers to obtain the signal. Most satellite receivers include key management applications that allow senders to encrypt the stream and share the key with the intended recipients of the signal. These signals may be used for a long period of time

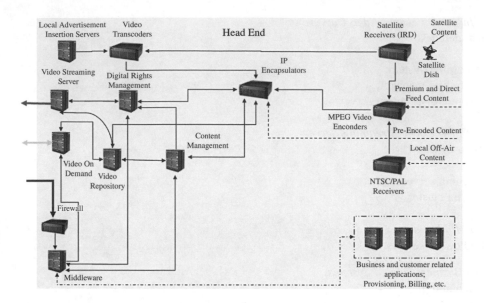

**Figure 6.2**  IPTV head end

**Table 6.1**  Key management countermeasures

| Key Management for Satellite Feed | |
| --- | --- |
| Theft/abuse of IPTV assets | This countermeasure would reduce the chances of theft of IPTV assets, mainly satellite feed being received by unauthorized parties. |
| Theft of service | Not applicable. |
| Theft of IPTV-related data | Not applicable. |
| Disruption of service | This countermeasure would reduce the chances of unauthorized modification or loss of the satellite receiver keys; if lost, the system could not be configured to receive the signal and there would be a temporary loss of content. |
| Privacy breach | Not applicable. |
| Compromise on platform integrity | Not applicable. |

and are valuable assets. Appropriate procedures should be deployed to avoid unauthorized access to the keys. Table 6.1 includes a reference to key management countermeasures.

### 6.2.2.2  Premium and Direct-feed Content, Pre-encoded Content Ready to be Encapsulated

In some cases, content will be received directly using VPNs, private networks, cable networks or other types of direct feed. This content is received by equipment that is not part of the IP network, and the best security is to ensure physical protections of the cabinet used for the equipment.

**Table 6.2**   Physical media countermeasures

| Physical Media Protections | |
| --- | --- |
| Theft/abuse of IPTV assets | This countermeasure would reduce the chances of theft of IPTV assets, mainly original content held on physical media. |
| Theft of service | Not applicable. |
| Theft of IPTV-related data | Not applicable. |
| Disruption of service | This countermeasure would reduce the chances of a disruption to the IPTV service owing to content being lost. |
| Privacy breach | Not applicable. |
| Compromise on platform integrity | This countermeasure would reduce the chances of unauthorized content being fed into the IPTV environment, as well as modifications to content. |

#### 6.2.2.3 Physical Media

A common mechanism to provide contents is the use of physical media. When the video content has to be modified by a third party, including subtitles and translation/doubling, the intermediary will deliver physical media. Physical media must be transported using vetted couriers and safe containers that do not allow tampering. The data must be encrypted to avoid unauthorized access to the contents while the media are in transit. Table 6.2 includes references to applicable physical media countermeasures.

### 6.2.3 MPEG Video Encoder and Video Transcoder Functions

The MPEG encoder is the first element that manages digital video. This component receives inputs from the NTSC/PAL receivers (off-the-air content), premium feed/content (VPNs, direct link, cable) and inputs from the satellite receivers. Up to that moment, the content has not been encoded, and any attempts to steal the content would require a very large container or a prolonged connection to the system.

MPEG encoders would require a connection to the network to allow for administrative and support activities. This connection can be abused at the operating system or application. The operating system of the MPEG encoder must be properly patched to avoid common security exploits and viruses. Additionally, the network access should be controlled to avoid unauthorized access to this system. VLANs can be set up for the MPEG encoder and transcoder to be able to communicate with the IP encapsulator on a dedicated VLAN, leaving another VLAN for administrative functions. In some cases the NTSC/PAL receivers and satellite receivers would be on the same computer rack as the MPEG video encoders. A direct connection would be possible without using the network.

Servers hosting the encoder and transcoder applications must be configured to remove any removable media or USB connector. This will reduce the chances of removal of digital assets. Table 6.3 includes references to MPEG and video transcoder countermeasures.

**Table 6.3**   MPEG and video transcoder countermeasures

| MPEG and Video Transcoder Physical Access Restrictions | |
| --- | --- |
| Theft/abuse of IPTV assets | This countermeasure would reduce the chances of theft of IPTV assets, mainly MPEG encoded assets. |
| Theft of service | Not applicable. |
| Theft of IPTV-related data | Not applicable. |
| Disruption of service | This countermeasure would reduce the chances of a disruption to the IPTV service owing to the MPEG application not being available. |
| Privacy breach | Not applicable. |
| Compromise on platform integrity | This countermeasure would reduce the chances of unauthorized content being fed into the IPTV environment, as well as modifications to content. |

## 6.2.4 IP Encapsulator

The IP encapsulator receives a feed from the MPEG encoders and the transcoder as well. This information is in digital form and easy to modify or steal. The IP encapsulator simplifies this further by creating an IP-ready content, making it easier and faster to stream over an IP network.

This system would have a connection to the local administrative network and the MPEG encoders and transcoders. The VLAN can be configured to allow the IP encapsulator to receive a feed from the MEPG and transcoder systems, and also to allow the IP encapsulator to communicate with the DRM, content management and video repository. The VLANs can include access control lists, filtering the source and destination of communications, allowing only approved systems to interact with the IP encapsulator. All administrative access should be done using secure communication channels (such as SSH, TLS, SSL and SNMPv3).

Access controls must be implemented to allow only authorized access to the system. These include authentication, identification and authorization of users and systems having access to the IP encapsulator.

Servers hosting the IP encapsulator application must be configured to remove any removable media or USB connector. This will reduce the chances of removal of digital assets. Table 6.4 includes references to IP encapsulator countermeasures.

In general, most servers within the head end are protected by six security layers formed mostly by independent mechanisms. These layers tend to comprise different technologies and be administered by different teams within the unit. This increases the controls and reduces the chances of collusion and fraud. Figure 6.3 illustrates the different layers.

The outermost layer is formed by the network firewalls and network IDS/IPS elements. These will provide access controls between the main functions and exchanges, as well as protection against known attacks such as worms and viruses.

The second layer comprises the VLAN established at the switches. The VLAN is configured to allow only preapproved hosts to join the network. Servers will have one network card dedicated to the administrative VLAN and several NIC cards for other VLANs according to the required communication needs. Access to the VLAN is controlled on the basis of the MAC address of the NIC or the IP address of the host. In some cases, servers can be authenticated using credentials before joining a VLAN.

The third layer is formed by the access control list at the switch. In most cases, communications will be initiated by one host and accepted by the destination host. This

**Table 6.4** IP encapsulator countermeasures

| IP Encapsulator Layered Access Controls | |
|---|---|
| Theft/abuse of IPTV assets | This countermeasure would reduce the chances of theft of IPTV assets, mainly IP encapsulated streams. |
| Theft of service | Not applicable. |
| Theft of IPTV-related data | Not applicable. |
| Disruption of service | This countermeasure would reduce the chances of a disruption to the IPTV service owing to the IP encapsulation function not being available. |
| Privacy breach | Not applicable. |
| Compromise on platform integrity | This countermeasure would reduce the chances of unauthorized content being fed into the IPTV environment, as well as modifications to content. |

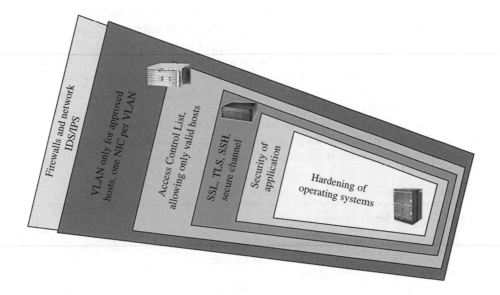

**Figure 6.3** Security layers at the head end

traffic will flow in preapproved directions and to specific ports and services. For example, the IP encapsulation server will not be able to send packets to the SMTP (Simple Mail Transfer Protocol) port of the content management server. Administrative consoles will have more flexible access to servers, but it will be validated by the firewall.

The fourth layer is formed by the encapsulation provided by the secure protocols. This is basically a tunnel established between hosts using SSL, TLS, SSH, SNMPv3 or similar secure communication channels that would protect the session from interception and modification as well as reducing the chances of a man-in-the-middle attack.

The fifth layer is formed by the specific security mechanisms implemented by the software vendor who created the particular application. In most cases, applications will request credentials before allowing access as user or administrator and will assign different profiles to users.

The sixth layer is the hardening of the platform, which includes the patching and configuration of the operating system, in some cases following the recommendations from the SCAP as mentioned at the beginning of the chapter.

With this layered approach it becomes very difficult to take control of elements within the head end. However, these rules do not apply to customers facing servers such as middleware or VOD servers, which will be discussed later in the chapter.

## 6.2.5 Content Management Server

The content management server is one of the key components within the IPTV environment. It does not hold any digital assets, but it provides interaction and control over the flow of the streams. Any damage to this component would affect the operation of the IPTV service owing to lack of contents, or may facilitate unauthorized access to contents.

The content management server must be part of a dedicated VLAN where only authorized systems are allowed to interact. A particular network interface could be dedicated to communication with peer systems; for example, one network interface for the VLAN with the administrative console, another for communication with the IP encapsulator, one card for the VLAN with the video repository, etc. With this approach, the VLAN will provide a secure environment where only the expected systems are exchanging information. Additionally, the access control list would allow only authorized flows of information. All this will be supported by ensuring that communications occur encapsulated on secure communication channels, for example SSL, TLS, SSH or SNMPv3.

All metadata assigned to the digital assets should be protected using validation mechanisms such as checksums or digital signatures. This will avoid unauthorized modifications. Table 6.5 includes references to content management server countermeasures.

**Table 6.5**  Content management server countermeasures

| Content Management Server Layered Access Controls | |
| --- | --- |
| Theft/abuse of IPTV assets | This countermeasure would reduce the chances of theft of IPTV assets, mainly redirection of IPTV content. |
| Theft of service | Not applicable. |
| Theft of IPTV-related data | This countermeasure would reduce the chances of theft of IPTV metadata linked with IPTV assets. |
| Disruption of service | This countermeasure would reduce the chances of a disruption to the IPTV service owing to the content management server being out of service. |
| Privacy breach | Not applicable. |
| Compromise on platform integrity | This countermeasure would reduce the chances of unauthorized content being fed into the IPTV environment, as well as modifications to content. |

## 6.2.6 Video Repository

The video repository is clearly the most valuable host within the IPTV environment owing to the volume of data stored. Ideally, access to this server must be limited to only required hosts. In reality, many servers must be allowed to communicate with this host, including the

DRM server, IP encapsulator, content management server, video on demand and in some cases the video streaming server.

This server requires additional hard-drive encryption to ensure that video assets are properly protected even in cases where administrators, technical support personnel or intruders are attempting to extract digital contents. In some cases, hard drives fail and need to be disposed of. Managing hard-drive encryption will help to protect contents once the disk has been destroyed according to the disposal policies.

Servers hosting the video repository application must be configured to remove any removable media or USB connector, which will reduce the chances of removal of digital assets.

All metadata assigned to the digital assets should be protected using validation mechanisms such as checksums or digital signatures. This will avoid unauthorized modifications. Table 6.6 includes references to video repository countermeasures and Table 6.7 to video repository disk countermeasures.

**Table 6.6**   Video repository countermeasures

| Video Repository Layered Access Controls | |
| --- | --- |
| Theft/abuse of IPTV assets | This countermeasure would reduce the chances of theft of IPTV assets, mainly theft of IPTV assets. |
| Theft of service | Not applicable. |
| Theft of IPTV-related data | This countermeasure would reduce the chances of theft of IPTV metadata linked with IPTV assets. |
| Disruption of service | This countermeasure would reduce the chances of a disruption to the IPTV service owing to contents being destroyed or modified. |
| Privacy breach | Not applicable. |
| Compromise on platform integrity | This countermeasure would reduce the chances of unauthorized content being fed into the IPTV environment, as well as modifications to content. |

**Table 6.7**   Video repository disk countermeasures [2]

| Video Repository Disk Encryption | |
| --- | --- |
| Theft/abuse of IPTV assets | This countermeasure would reduce the chances of theft of IPTV assets, mainly theft of stored video. |
| Theft of service | Not applicable. |
| Theft of IPTV-related data | This countermeasure would reduce the chances of theft of IPTV metadata linked with IPTV assets. |
| Disruption of service | This countermeasure would reduce the chances of a disruption to the IPTV service owing to the content being modified or removed. |
| Privacy breach | Not applicable. |
| Compromise on platform integrity | This countermeasure would reduce the chances of unauthorized content being fed into the IPTV environment, as well as modifications to content. |

## 6.2.7 Digital Rights Management

The DRM is a key component required to ensure protection of the digital assets being provided by the IPTV service. In some implementations of DRM, the set top box would be allowed to connect to the DRM server. This creates a security problem that must be controlled to avoid future security incidents. The previously mentioned six layers of security are not applicable to the DRM service when subscribers are allowed to connect. In that case, the firewalls and VLANs would include access to an URL from where set top boxes can download the keys to decrypt content. The security layers are opened for that type of access as illustrated in Figure 6.4.

The Firewall would allow access by all set top boxes, specifically connections to port 80/443 on the DRM server. The VLAN will be open to all set top boxes interested in requesting keys, which provides a very large number of potential attackers. Filters can be deployed at the switch level, filtering connections to ports different from 80/443 coming from set top boxes. Connections will be made using SSL, but this does not provide any protection if the attacker is trying to exploit a vulnerability within the web services. Web application vulnerabilities may allow intruders to take control of the server even if other mechanisms have been deployed.

For servers allowing access by set top boxes to web services there are three main countermeasures that may be deployed: web services gateway, reverse proxy or web application firewall.

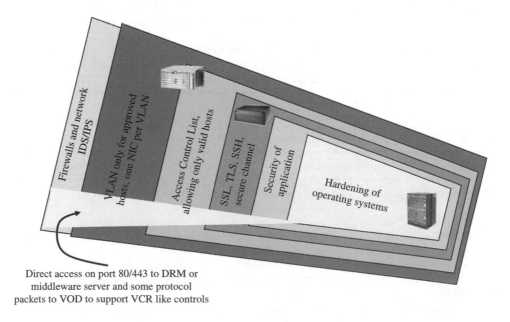

Direct access on port 80/443 to DRM or middleware server and some protocol packets to VOD to support VCR like controls

**Figure 6.4** Security layers – direct access to DRM

*(i) Web Services Gateway (WSG)*
This is part of the service-oriented architecture (SOA) approach and allows architects to integrate all web services under a single front with unified authentication, security and auditing functions. If a security incident occurs, it will be restricted to the WSG and will not immediately affect the web servers. This adds time to facilitate the detection and containment of the attack. The DRM server can have one NIC dedicated to communications with the WSG, again allowing the six security layers.

*(ii) Reverse Proxy*
This component is an intermediary between the set top boxes and the DRM server. The reverse proxy terminates the session with the set top box and establishes another session with the DRM server. The DRM can have one NIC dedicated to communications with the reverse proxy, thus reducing dramatically the number of hosts allowed to connect with the DRM server and again allowing the six security layers.

Reverse proxies would check the validity of the HTTP requests and confirm if they are compliant with the protocol. This will filter out most buffer overflows and DOS attacks. Authentication and authorization functions will allow only connections with valid addresses within the DRM server. Understanding that there would be one or two valid folders/pages within the DRM site, it is easy to assume that intruders will not be able to exploit attacks against other files/pages/areas of the web server. This eliminates the default vulnerabilities found on the sample pages and basic configuration vulnerabilities within the web server.

*(iii) Web Application Firewall*
The web application firewall blocks all access to unauthorized pages within the DRM web site. It would also block requests that do not conform to preapproved values or structures. Responses will also be prevalidated, and any abnormal response will be blocked. One example would be intruders sending an SQL injection attack against the server. This would be a request using characters that are not part of the preapproved structure. The web application firewall will block the request and no packets will be forwarded to the DRM server; even if an attack were allowed to pass, the response of an SQL injection would not conform with the standard DRM response, and hence it would be blocked.

The DRM will have a dedicated NIC to communicate with the web application firewall, allowing the six protection layers. Table 6.8 includes references to DRM layered countermeasures.

The three solutions for web services security (WSG, reverse proxy, web application firewall) are applicable to an IPTV environment where each deployment would have different requirements and budgets available. Table 6.9 includes references to DRM web service countermeasures.

Private keys from the DRM service must be protected at all times, as they are used constantly. The server must have mechanisms to protect the DRM service private keys. Additionally, all critical data related to the set top boxes and symmetric keys used for encrypting video streams must be protected. There are external cryptographic storage devices as well as disk encryption technologies that can be used. When choosing the solution, this must be in accordance with the value of the assets to be protected and the exposure levels.

**Table 6.8**   DRM layered countermeasures

| DRM Layered Access Controls | |
| --- | --- |
| Theft/abuse of IPTV assets | This countermeasure would reduce the chances of theft of IPTV assets, mainly theft of IPTV assets through stolen DRM keys. |
| Theft of service | Not applicable. |
| Theft of IPTV-related data | This countermeasure would reduce the chances of theft of DRM keys. |
| Disruption of service | This countermeasure would reduce the chances of a disruption to the IPTV service owing to DRM keys not being available for set top boxes to decrypt content. |
| Privacy breach | Not applicable. |
| Compromise on platform integrity | This countermeasure would reduce the chances of DRM private keys being lost to intruders. |

**Table 6.9**   DRM web service countermeasures

| DRM - Web Service Security | |
| --- | --- |
| Theft/abuse of IPTV assets | This countermeasure would reduce the chances of theft of DRM keys. |
| Theft of service | Not applicable. |
| Theft of IPTV-related data | This countermeasure would reduce the chances of theft of DRM keys. |
| Disruption of service | This countermeasure would reduce the chances of a disruption to the IPTV service owing to DRM keys not being available for set top boxes to decrypt content. |
| Privacy breach | Not applicable. |
| Compromise on platform integrity | This countermeasure would reduce the chances of DRM private keys being lost to intruders. |

This encryption must work even in cases where the server administrator, network administrator or any individual authorized to access the server but not the keys attempts to retrieve the keys. Keys should be accessed only by the DRM application, not by administrators or users. If the disk encryption relies on the operating system, it may not be a viable solution as an administrator may be able to manipulate the operating system to retrieve the keys.

The DRM as well as the middleware and VOD authentication must have additional mechanisms to prevent unauthorized access, specifically related to brute force.

In sections where subscribers are expected to enter personal identification numbers (PINs) or passwords, the system must include random pauses when checking for password information. This will cause problems to automated scripts used for brute-force attacks. The system must also record the IP address and subscriber identification used for failed attempts and lock IP addresses after a high number of failed requests have been logged.

**Table 6.10**   DRM disk countermeasures

| DRM Disk Encryption | |
| --- | --- |
| Theft/abuse of IPTV assets | This countermeasure would reduce the chances of theft of DRM keys that could end up facilitating abuse of IPTV assets. |
| Theft of service | Not applicable. |
| Theft of IPTV-related data | This countermeasure would reduce the chances of theft of DRM keys. |
| Disruption of service | This countermeasure would reduce the chances of a disruption to the IPTV service owing to the DRM keys being removed or modified. |
| Privacy breach | Not applicable. |
| Compromise on platform integrity | This countermeasure would reduce the chances of modification to the keys. |

For high-risk applications, and specifically for online access to the network PVR, it is recommended that CAPTCHA™ (Completely Automated Public Turing test to tell Computers and Humans Apart) be used [3]. This works as a challenge–response test mechanism that presents images and expects characters to be provided as a response. Automatic software is not able to read the text on the images, as it is obfuscated and will tend to fail. Table 6.10 includes references to DRM disk countermeasures.

## 6.2.8 Video Streaming Server

The video streaming server receives encrypted content from either the video repository or directly from the DRM in encrypted form. A dedicated VLAN can be set up for these elements to communicate. With this approach, the VLAN will provide a secure environment where only the expected systems are exchanging information. Additionally, the access control list would allow only authorized flows of information. All this will be supported by ensuring that communications occur encapsulated in secure communication channels, for example SSL, TLS, SSH and SNMPv3.

The video repository will also participate in the TV broadcast VLAN. This VLAN will include multicast traffic with all the TV channels available. The main security advantage is that multicast traffic does not need bidirectional communication between set top boxes and the video streaming server(s). The VLAN can be configured to block any traffic originated from the set top box network. The possibilities of attack to the video streaming server are very limited. This server will have the standard hardening and security layers implemented, and, facing the home end, the VLAN will be filtered at different levels (residential gateway, DSLAM, aggregation and switch level).

In some cases the video streaming server will send files to be stored by the set top box. It is important to follow these recommendations:

- Disable anonymous FTP log-ins.
- SFTP (i.e. secure FTP) should be used instead of FTP. SFTP does not provide authentication and integrity. The underlying protocol must provide authentication and

integrity. SFTP is most often used as a subsystem of SSHv2, which provides integrity protection and authentication.
• Configure ACLs on firewall interfaces to block SFTP packets originating from improper interfaces or with improper source IP addresses.

### 6.2.8.1 IGMPv2/v3

Multicast traffic originating from the head end and encapsulated within the VLANs must be protected from unauthorized modifications. An alternative, which is recommended for these two standards, is the use of an IPSec authentication header. Deploying IPSec protection would provide authentication and integrity protection.

The IPSec authentication header (AH) is used to provide connectionless integrity and data origin authentication for packets, reducing the exposure to replay attacks. AH provides for the IP header; some fields may change during transport.

IPsec AH can be used in conjunction with IP encapsulating security payload (ESP). ESP provides payload protection.

The different firewalls can be configured to block any IGMPv2/v3 traffic that is not originating from authorized broadcast addresses.

To avoid DOS attacks, it is recommended that IGMPv3 be used, as IGMPv2 does not suppress reports when receiving a membership report.

All set top boxes and residential gateways receiving IGMP packets should verify and discard them if the Ethernet MAC address is not a multicast Ethernet address [4].

### 6.2.8.2 MBGP

It is recommended that MBGP authentication and integrity parameters be enabled, and that ACLs be configured on firewall interfaces to block MBGP messages originating from improper interfaces or with improper source IP addresses.

### 6.2.8.3 MSDP

It is important to implement PIM-SSM (source-specific multicast) [5].

Security would be improved by enabling MSDP authentication/integrity parameter (MD5) [6].

It is recommended that ACLs be configured on firewall interfaces to block MSDP messages originating from improper interfaces or with improper source IP addresses (e.g. set top boxes cannot be multicast sources).

### 6.2.8.4 RTP

To reduce the exposure to forged or modified RTP packets [7], SRTP should be used instead of RTP, as it provides integrity protection and authentication in the form of HMAC-SHA-1 for the RTP and RTCP packets. The secure real-time transport protocol (SRTP) specification provides confidentiality via encryption, message authentication and replay protection for RTP and RTCP.

It is recommended that ACLs be configured on firewall interfaces to block RTP/SRTP packets originating from improper interfaces or with improper source IP addresses.

Additionally, to avoid the possibility of third-party capture of RTP packets in transit, SRTP should be used instead of RTP, as it provides encryption in the form of AES for the RTP and RTCP packets.

### 6.2.8.5 RTSP Packets

To avoid forged or unauthorized RTSP packets, it is recommended that RTSP basic authentication (shared secret) or digest authentication (MD5) be enabled on appropriate VOD server and STB interfaces. Additionally, it is important to protect RTSP packets with Ismacryp (i.e. ISMA 1.1). Ismacryp uses SRTP to provide transport authentication, integrity and encryption.

For additional protection against DOS and unauthorized access it is important to configure ACLs on firewall interfaces to block RTSP messages transporting RTSP packets originating from improper interfaces or with improper source IP addresses.

### 6.2.8.6 RSVP

The resource reservation protocol (RSVP) is used to reserve resources within a multicast environment. RSVP allows network elements to request access to resources. The assignment of permission depends on the availability and policies. However, this could be abused to cause network problems. RSVP requires protection mechanisms including authentication digests of messages based on secret authentication keys and a keyed-hash algorithm. This digest helps to reduce the possibilities of modifications in transit and replay attacks.

It is recommended that ACLs be configured on firewall interfaces to block RSVP messages originating from improper interfaces or with improper source IP addresses [8–10].

Additional information on RSVP can be found in the following Internet Engineering Task Force documents: RFC 2236, 3376, 4601, 2747.

## 6.2.9 Middleware Server

The middleware servers are the front end of the IPTV environment. All set top boxes are allowed to communicate with the middleware server to request the specific content they require. In some cases the middleware server will also provide DRM keys and may also receive requests for VOD content that will be forwarded to the VOD server.

The communication between the set top box and the middleware server is usually done using HTTP (HTTPS, SSL, TLS). A browser within the set top box will communicate with the middleware server, requesting the specific information.

This level of access by set top boxes directly to the middleware presents a security problem. Direct access to the middleware reduces the protection presented by the security layers. As explained in the DRM section, set top boxes would be able to communicate across the firewall, VLAN, ACL and in some cases even the security of the application.

To avoid security incidents and unauthorized access to the middleware server, the web service security mechanisms must be deployed, as these countermeasures would thwart hacking attempts aimed at the middleware.

## 6.3 Aggregation and Transport Network

### 6.3.1 DSLAM

The DSLAM is, in reality, the first trustworthy line of defense for the IPTV environment. Set top boxes are deployed by the IPTV service provider, but they are physically outside the protection of the vendor. The residential gateway is also outside physical control of the IPTV service provider. There are many reported cases of cable modems and set top boxes being disassembled to extract information and then reassembled with modified parameters. This is used by hackers and attackers to lift any restrictions on the operation of the set top box and also in attempts at theft of service.

The responsibility for the protection against malicious and unauthorized users trying to access the IPTV environment tends to reside with the network security mechanisms. Specifically, when talking about the IPTV network utilizing broadband access, it is mainly a task for the IP DSLAM as the first reliable element in the infrastructure. DSLAM is the place where the physical subscriber connection (the copper or fiber cable itself) is available and directly linked. The access node is the only location where the subscriber identity can be linked to a network authentication protocol for future use by the rest of the infrastructure.

The authentication is mostly based on the DHCP option 82 protocol. The DHCP option 82 functionality snoops the DHCP request coming from the subscriber network and inserts the physical line identification information in the option 82 field of the DHCP. This information typically includes access node ID, shelf ID, slot ID and finally the line ID. Processed more deeply in the network by the DHCP server, this option is used to identify the subscriber and assign (or refuse to assign) a valid IP address for the subscribed services. This type of approach ensures that requests coming from a particular subscriber will always be linked with the physical location of the subscriber. Additionally, even if the subscriber has control over the set top box and the residential gateway, information will always be added by an additional component outside the control of the subscriber.

Next to the above-mentioned authentication there are security mechanisms used in DSLAMs. On layer 2 (L2) there are mechanisms such as MAC antispoofing, and on layer 3 (L3) there are mechanisms such as IP antispoofing.

The task of MAC antispoofing is to protect the network from a malicious user configuring and using (spoofing) the MAC addresses of other subscribers. This could result initially in theft of service, granting unaccounted access to services for the malicious user. A second effect could be denial of service towards paying subscribers, with the malicious user preventing the valid subscriber from accessing services. A strict MAC antispoofing prevents any user from using an already known MAC address. MAC addresses are linked with physical locations and subscribers, and adding one would require authentication by the subscriber.

Once the L2 security is granted, protection mechanisms must be implemented on L3 as well. In a DHCP-enabled broadband network, malicious users could connect a PC instead of their STB and manually configure any static IP address (for instance that of another subscriber). Without any protection, this would result in granting the malicious user access to the regular subscriber's services, without being accounted for. With IP antisnooping, the access node blocks all traffic from the subscriber (both downstream and upstream) before a DHCP request and a trusted DHCP reply from the DHCP server are sent, meaning that a non-DHCP-authenticated subscriber cannot access any service or impact on any other subscriber. Once this assignment is granted, the IP-aware bridge instance strictly binds the subscriber's trusted IP address with the DSL physical line, and only allows traffic associated with these trusted addresses. This mechanism is fully automatic and learns the IP address by snooping the DHCP protocol, avoiding the need for any static configuration.

Other protection functionality used in DSLAMs is user-to-user communication blocking. At an access node level, this feature offers to prevent any L2 (e.g. Ethernet) user-to-user traffic exchange. This guarantees a strict separation of each subscriber home network and prevents low-level network attacks on subscriber equipment. Other protection mechanisms could be the virtual MAC address (MAC address translation by DSLAM) and, last but not least, filtering – a policy based on a wide set of rules, controlling traffic to and from DSL subscribers.

Adding these DSLAM protection mechanisms, it is possible to ensure that any communication accepted comes from a pre-established and authorized physical node and there are no options for cross-communication between users. Worms, viruses and hackers will have access only to valid nodes upstream, and those nodes would be protected.

The DSLAM is held within a protected physical environment where the logical configuration can be protected and ensured. This is the IPTV service provider's primary line of defense.

There are three main functions within a DSLAM:

1. *IP concentration.* This function covers the concentration of data received by the other functions and the preparation of data for transmission.
2. *IP services.* This function covers the advanced IP functionality and includes areas such as routing and bridging. L2 and L3 capabilities are found within this function. Most of the security mechanisms available within the DSLAM are found here.
3. *DSL line service.* This function covers the physical link with the customer's premise termination equipment.

Most of the DSLAM security characteristics reside with the IP services function. The key aspects to consider are:

- access and session control;
- routing;
- user segregation;
- quality of service (QoS);
- virtual networks and virtual service;
- 802.1X authentication.

### 6.3.1.1 Access and Session Control

The DSLAM supports a series of subscriber authentication and session tracking capabilities that are critical for validating set top boxes within the IPTV environment. Using credentials stored on their set top boxes, subscribers are authenticated against a local user database stored on the DSLAm or through an external RADIUS server. The DSLAM can exchange the authentication data with the RADIUS server or even with the middleware server. Subscriber and physical link data can be considered during the authentication process. Only authenticated physical ports are allowed to exchange information with the transport network and head end. This mechanism greatly reduces the chances of unauthorized access to the network. There is a one-to-one relationship between the physical ports and the subscribers. Any attacker must authenticate and use a valid physical port.

DSLAMs can validate and authorize users on the basis of the MAC address used at the other side of the physical link, as well as other common authentication mechanisms such as challenge handshake authentication protocol, IP address authentication and PPP authentication protocol. As ports are linked with the MAC address of the subscriber, it is not possible to impersonate another subscriber. Within the IP world it is relatively easy to send spoofed messages or hijack a valid IP session. However, the validation provided by the DSLAM eliminates this type of attack.

IPTV environments include thousands of subscribers. This represents between 5000 and 10 000 subscribers per DSLAM, and the only practical way of managing IP addresses would be to use dynamic host configuration protocol (DHCP). When DHCP is used to assign IP addresses, the DSLAM can use DHCP option 82 to insert data related to the physical link. The authentication of set top boxes is primarily based on the DHCP option 82 protocol. DHCP option 82 snoops the DHCP request coming from the subscriber network and inserts the physical line identification information in the option 82 field of the DHCP package. This information typically includes access node ID, shelf ID, slot ID and finally the line ID. Processed more deeply in the network by the DHCP server, this option is used to identify the subscriber and assign (or refuse to assign) a valid IP address for the subscribed services.

If a subscriber attempts to bypass the authentication provided by the DSLAM, the system easily recognizes that the subscriber has manipulated parameters. For example, suppose user A attempts to spoof a valid MAC address from subscriber B by modifying his or her own MAC address and replacing it with the one from B. In this case, the DHCP server will identify that the MAC address submitted by subscriber A during the DHCP initial request corresponds to a MAC address already registered with a different physical location (the physical location of B). Using this process, physical line information is linked with the MAC address of the subscriber and spoofing and hijacking are not possible.

DHCP option 82 is a powerful mechanism that can be used to ensure tighter control over access to the infrastructure. The approved MAC addresses can be stored with the physical information, allowing each subscriber to have a small number of network elements preauthorized to access the network. Each time a new MAC (network element) attempts to join the network, the system may ask for authentication information (a user name and password) before adding that particular MAC to the database. Properly deployed, DHCP option 82 will eliminate most opportunities for MAC and IP spoofing. It will also increase the level of confidence with regard to system information, and also reduces the incidence, frequency and impact of unauthorized access to content.

DSLAMs can also provide access control lists (ACLs) and usage restrictions to control the type and amount of data sent by subscribers. The DSLAM can provide additional filtering to the rules provided by the residential gateway. More specifically, the DSLAM can control the type of requests sent by the set top box over the different VLANS. For example, the DSLAM can be configured to allow only HTTP, HTTPS, DHCP, DNS and RTSP to flow upstream. By dropping all requests to unnecessary ports, the DSLAM protects the rest of the infrastructure from DOS attacks coming from set top boxes, and it also reduces the possibilities of a distributed DOS. Table 6.11 includes references on how access control lists can be used to ensure confidentiality, integrity and availability.

**Table 6.11**   How ACL security ensures confidentiality, integrity and availability

|  | Confidentiality | Integrity | Availability |
|---|---|---|---|
| Access and session Control | DHCP option 82 reduces the chance of MAC and IP spoofing. | Attackers are not able to spoof a valid address in order to attack the head end and modify information. | Attackers are not able to spoof a valid address in order to launch DOS attacks. |

### 6.3.1.2 Routing

Service providers can use L3 static, dynamic and policy-based routing. Policy-based routing is implemented using several virtual routing domains (VRDs). Each VRD is used to provide the routing decision for subscriber packets and is based on the routing entries of the VRD to which the subscriber belongs.

With the basic routing, packets can be segmented within valid domains only, ensuring that only authorized network elements transmit within the IPTV VRD. Unauthorized set top boxes will not be able to send packets within the IPTV network. This will include cases where intruders try to enter the network by hijacking a physical link.

Routing can be used to provide mechanisms that protect confidentiality, integrity and availability. Table 6.12 presents a view on how routing can be used to ensure confidentiality, integrity and availability.

### 6.3.1.3 User Segregation

With the imminent threat from computer viruses and worms, along with the clear risk of subscribers breaking into their neighbors' set top boxes, it is necessary to segregate users within their own networks. Subscriber segregation ensures that subscribers are not able to access each other's set top boxes, reducing the impact of computer worms and viruses. If an intruder takes control of one set top box, physically or exploiting a vulnerability over the Internet, no packets would be allowed between that set top box and neighboring set top boxes. This is particularly useful to reduce the impact of worms and viruses that use algorithms to guess the IP addresses of neighboring boxes to infect them with copies of the virus. As the traffic is not allowed, it would be dropped by the DSLAM without affecting the upstream link. Cross-talk between subscribers is blocked, and only valid traffic is sent to the aggregation elements. Routing rules will not allow cross-talk, as it is not a service provided to subscribers.

**Table 6.12**   How routing ensures confidentiality, integrity and availability

|  | Confidentiality | Integrity | Availability |
|---|---|---|---|
| Routing | Creating VRDs reduces the risk of unauthorized access to information. | Within the VRDs, information is maintained without modifications. | Each VRD is assigned specific throughput and QoS characteristics. |

**Table 6.13** How user segregation ensures confidentiality, integrity and availability

|  | Confidentiality | Integrity | Availability |
|---|---|---|---|
| User segregation | STB and other equipment at the customer premises is secure from access by other subscribers | Intruders are not able to cross talk with other STB and manipulate their information. | Attackers are not able to spoof launch DOS attacks against other STB and customer equipment |

Users are allowed to access only the head end for IPTV services. No other requests will be allowed. Table 6.13 presents a view on how user segregation can be used to ensure confidentiality, integrity and availability.

### 6.3.1.4 Quality of Service

The DSLAM is capable of classifying traffic to ensure that QoS is maintained at all times. The DSLAM enforces rules to guarantee that critical services are always available. This provides additional protection against denial of service attacks by subscribers trying to use all the bandwidth available for HTTP access to the head end. Sessions are controlled to ensure that there is bandwidth available for critical services. While designing the DSLAM and upstream bandwidth requirements, architects would be able to define the maximum bandwidth allowed for each set top box. Any attempts to exceed this allocated bandwidth will cause the QoS mechanism to drop the additional traffic.

Within the IPTV environment it is critical to ensure that subscribers are not able to flood the platform by sending too many requests to the head end. A minimum bandwidth must be guaranteed to ensure that subscribers are able to connect to the head end at all times to request services. Additionally, the broadcast content sent by the head end requires minimum bandwidth levels to ensure customer satisfaction. Table 6.14 shows how quality of service can be used to ensure availability.

### 6.3.1.5 Virtual Networks and Virtual Circuits

When virtual private networks (VPNs and VLANs) are used to transmit valuable information using a public network, data is protected from unauthorized access.

**Table 6.14** How QoS ensures availability

|  | Confidentiality | Integrity | Availability |
|---|---|---|---|
| Quality of service | — | — | Attackers will not be able to use all the bandwidth, causing a DOS. |

**Table 6.15** How user virtual networks and virtual circuits ensure confidentiality, integrity and availability

|  | Confidentiality | Integrity | Availability |
|---|---|---|---|
| Virtual networks and virtual circuits | Information is segmented, allowing better protection of confidentiality of information. | There are clear restrictions on the type of traffic allowed, reducing the type of attacks that can affect the integrity of data and systems. | Traffic is segmented, reducing the possibilities of flooding. |

Within the VPN, a group of authorized network elements is able to access the content.

Virtual networks and virtual circuits can be used to segregate the IPTV traffic from all other traffic flowing through the DSLAM. With QoS and VLANs, the traffic can be controlled in a way ensuring that the IPTV VLAN has guaranteed bandwidth and service is maintained in spite of attempted DOS attacks.

VLAN and/or VPNs can be used to segregate the traffic between specific circuits dedicated to services such as IPTV, VoIP, Internet access, control, etc. Within each circuit, different rules and security mechanisms can be deployed, certain traffic can be blocked and the amount of noise within the network can be dramatically reduced. For example, in the VLAN dedicated to IPTV, security mechanisms can be deployed to ensure that only HTTPS requests are sent between the STB and the head end.

Depending on the type of equipment available, this function can be deployed using L2 or L3 filtering – these are linked to other security requirements and capabilities of the equipment. Table 6.15 explains how virtual networks ensure confidentiality, integrity and availability.

### 6.3.1.6 802.1X Authentication

The IEEE developed a standard for local and metropolitan area networks, specifically covering port-based network access control [11]. Many network environments have a number of physical ports that are exposed to unauthorized access. These physical ports are available for access and require logical controls to avoid security incidents. This situation is applicable to the IPTV environment, specifically for set top boxes and their connections to the DSLAM ports (via the residential gateway).

The purpose of the standard is to allow very restricted communication between a newly connected host and a validating host until the identity has been verified and access is granted. Connected devices are known as supplicants, intermediaries are known as authenticators (in this case the DSLAM) and there is an authentication server who would make the final decision (in this case the RADIUS server using EAP). The supplicant starts an 802.1X request using EAPoL. This is received by the authenticator and forwarded to the authentication server using EAP/RADIUS as part of the negotiation. Once the authentication server has verified and authorized the supplicant, then the port configuration

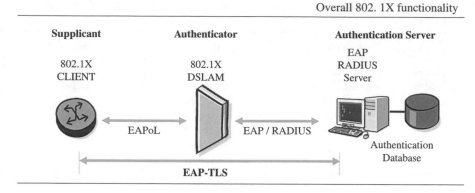

**Figure 6.5**   802.1X functionality

will be modified and restrictions will be lifted. The authentication process is illustrated in Figure 6.5.

In more detail, the exchange of information with the DSLAM and authentication server will be as follows:

• An initial EAPOL package is sent by the supplicant to start the process.
• The authenticator responds with an EAP package requesting the identity of the supplicant.
• The supplicant will respond with the identity information, which will be used by the DSLAM to get confirmation from the RADIUS server on the level of access expected from the supplicant.
• A challenge is issued by the RADIUS server and forwarded by the DSLAM to the supplicant.
• A challenge response is issued by the supplicant and forwarded by the DSLAM to the authentication server along with the access request. With a satisfactory validation, the authentication server will issue a package with port authorization. If the authentication fails, the port will not be authorized.
• Both responses are forwarded to the supplicant, and at the same time the relevant changes to the port are implemented.

Packet traffic is illustrated in Figure 6.6.

## 6.3.2 Firewalls

Network firewalls are required for protecting the traffic sent upstream from the DSLAM towards the middleware server. Even if the residential gateway and the DSLAM are blocking requests to any unauthorized port, there are some risks that intruders may be able to bypass those security mechanisms.

Before the home end, a network firewall must be deployed to filter all requests and allow only valid requests to the middleware server. The web server security mechanisms provide security protection to port 80/443, but other ports may still be vulnerable.

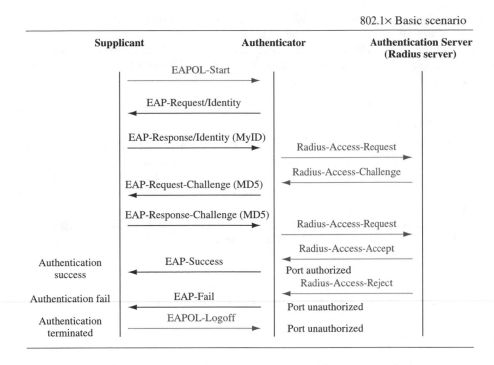

**Figure 6.6**  802.1X validation

## 6.4 Home End

The home-end environment presents some difficult challenges for the security of the IPTV environment. Components are outside the control of the IPTV service provider and hardware is exposed to modifications by attackers. Countermeasures deployed within the home end are intended to slow down attackers and block the most trivial attacks. Stronger security mechanisms are available at the aggregation network.

It is important to remember that the set top box has virtual access to the video repository via the standard valid requests. If intruders are able to take control of the set top box via the Internet, they may be able to extract digital contents from the set top box.

### 6.4.1 Residential Gateway

The residential gateway concentrates the different services provided to the subscriber into a single component. The residential gateway has some basic security mechanisms including filtering functions and QoS capabilities. The residential gateway shares different local connections including VoIP phones, high-speed Internet access and IPTV services.

#### 6.4.1.1 Filtering

The residential gateway can be configured to filter packets and allow only valid requests from the home end towards the head end. In particular, the filters can block any upstream traffic that does not conform to the expected traffic type (80, 443 and RSTP). This will reduce the chances of worms or computer viruses taking over large numbers of set top boxes to launch DOS attacks against the head end.

Filters are also able to block unauthorized access attempts from the head end towards the set top box and other equipment located at the home end. This will avoid port scans and similar attacks to the internal equipment.

#### 6.4.1.2 Quality of Service

Quality of service functions can be activated on the residential gateway to support the enforcement of QoS restrictions. Upstream traffic should require a relatively low bandwidth. If a set top box has been infected by a computer worm and is trying to send large amounts of data to valid ports (80, 443, RSTP), then the QoS will restrict the bandwidth and would limit the effect of the attack.

These QoS restrictions will be reinforced at different points between the home end and the head end, specifically at the DSLAM and firewalls.

### 6.4.2 Set Top Box

Set top boxes tend to be solid-state components that host all the functions in chips. There are some limited examples of small-form or PC-based set top boxes. Set top boxes share the same type of challenges, usually linked with the threats of unauthorized access to contents or encryption keys. Intruders will try to remove hardware components in order to understand how the set top box works, and in some cases will reprogram internal chips to embed additional software within the system. PCs allow for more simple access to the information and keys, as the attacker already has complete control over the operating system and the only additional activities required are trying to capture keys while they are in memory or in some cases trying to capture content after it has been decrypted. In general, set top boxes are exposed to attacks and must be considered a nonsecure device.

Set top boxes are designed to store the operating system and clients using hardware components. One of the main components within the operation of set top boxes is flash memory. Flash memory is nonvolatile memory that can be electrically erased and reprogrammed (it is a type of electrically erasable programmable read-only memory – EEPROM). These components can be used to store software code and keys. Information will be stored even if the set top box is disconnected from the energy source.

While designing the IPTV environment, security professionals must ensure that the set top box selected has an acceptable architecture and appropriate security characteristics, ensuring protection from attacks. When comparing different set top box models, security professionals must explore the security characteristics of each model and confirm how well each model will withstand a local/hardware intrusion.

The operating system used by the set top box must be hardened following the recommendations at the beginning of the chapter. Unnecessary ports and services must be removed to avoid unauthorized access to the system.

#### 6.4.2.1 Secure Processor

Some set top box manufacturers are able to include chip sets with security mechanisms to protect the programs within the equipment. This will allow for secure storage of information, logic and routines. Intruders will not be able to capture information while it is stored in memory.

In some available models, any block of a flash device can be individually protected against illegal program or erase operations. Additionally, blocks can be locked so no future modification is allowed. Blocking sections of the flash memory ensure that intruders will not be able to reprogram the application. This will protect the set top box from insertion of backdoors and removal of protections and in general will ensure that the application will work as originally designed.

Other mechanisms available include authentication between the flash memory component and the CPU. Each element is able to validate its counterpart and also to provide credentials for verification. This type of authentication prevents illegal operations via unauthorized processor or flash memory connected in parallel. Attackers tend to replace elements that provide security barriers, and, in the case of flash memory with protected segments, intruders will try to replace it for a new chip without restrictions.

There are read protection mechanisms preventing unauthorized reading of memory, or duplication of data in pirate devices, hence safeguarding IP (intellectual property) and stored program code. Intruders will only be able to access encrypted memory entries. This will not facilitate the process of capturing decryption keys or even content. Additionally, application code will be secure from eavesdropping, adding complexity to the process of understanding how the code works.

In general, components of this type are referred to as secure system-on-chip processors. These elements can support antitamper and intrusion prevention. They have the ability to execute encrypted programs, and also to protect both data and code from intruders. All communication with the rest of the components is done through a secure channel. Secure processors comprise:

- processors;
- tamper detection system;
- key storage section;
- boot protection information;
- access controls;
- cryptographic engines;
- secure channel.

The tamper detection system is the overarching mechanism that will detect any attempt to manipulate the hardware. If the chip is removed from the board or there is an attempt to have unauthorized physical access to the component, the element will be damaged and operation will not be possible.

Secure channels are used to communicate in and out of the secure processor. Encrypted data is fed to the element, and output will also be protected.

Decryption and private keys are stored and used only within the elements of the secure processor. If the DRM or middleware servers send a symmetric key encrypted with the public key of the set top box, this packet will be received by the set top box in encrypted form and will be forwarded to the secure processor where it will be decrypted and stored for later use.

The cryptographic engines are used to accelerate the encryption-related processes and ensure a safe environment for the recovery of information. Encrypted content is sent to the secure processor via the secure channel and is decrypted using the available symmetric keys. The keys never leave the secure processor in clear text, and it is not possible for attackers to capture clear text information from memory.

The main purpose of the secure processor is to protect critical data. Encrypted code is received, decrypted and executed by the secure processor. Memory entries are encrypted, as well as data managed by the processor. There is no opportunity for critical data to be captured in open text form; all entries are encrypted.

### 6.4.2.2 DRM

DRM clients are responsible for the key negotiation and validation. Within PC environments, DRM clients have limited options to protect keys, as in most cases they would have to be exposed in memory.

The DRM client also participates in the PKI validation exchange, which includes presenting the set top box digital certificate, encrypting validation information (via the secure processor) and validating the credentials from the DRM server and the middleware server.

The DRM client and web browser must be configured properly to validate the PKI-related data. The certificate revocation list should be downloaded and checked frequently by the system. This is important to ensure that the CA and SubCA have not been revoked, as well as to check that DRM, middleware and VOD hosts have valid certificates. The CRL must be digitally signed by a valid CA, but the root CA certificates must be stored in a safe location within the STB and modification of these entries must be prevented. This will prevent modifications of the CA certificate. Intruders may want to plant fake CA certificates to trick the set top box into accepting a false DRM or middleware certificates.

In a similar situation, the digital certificate from the set top box and the private keys must be stored in a secure location within the set top box to avoid tampering. A copy of the credentials can be maintained on a smart card inserted in the set top box. This will present an additional layer of security.

### 6.4.2.3 Output Protection

Attackers may try to take control of set top boxes to extract valuable content streams. The set top box output may be intercepted in order to distribute copies of digital assets. It is important that all output be protected following international standards, examples of which are as follows [12, 13]:

- *High-bandwidth digital content protection* [14]. High-bandwidth digital content protection (HDCP) is used to protect digital assets from unauthorized reproduction. This standard covers digital output from the digital visual interface (DVI) and high-definition multimedia interface (HDMI). HDCP provides the necessary authentication mechanisms to block any high-definition output to unauthorized devices. The principle is that only authorized devices (which will not duplicate the content) are allowed to receive the content. The encryption of the content prevents interception or modification while transmitted. If a particular model/brand is known to have been hacked, then their keys can be revoked so no HD content is provided in the future to those devices. HDCP works by providing a set of 40 unique keys to each device. The keys are 56 bits in length.

An additional key selection vector (KSV) is assigned to the device and is used by the receiver and the set top box to exchange validation information and select the HDCP encryption keys.

- *Digital transmission content protection.* Transmission content protection (DTCP) is used to allow interconnection of devices at the home end. This allows for set top boxes, personal computers, media consoles and other devices to be connected using USB, PCI, Bluetooth, Firewire and IP. The set top box output will be limited to DTCP authenticated components. This will prevent unauthorized copies of digital contents.

Some set top box manufacturers are starting to add watermarking and fingerprint chip sets to their products. This technology allows IPTV service providers to embed information about the subscriber with the video stream leaving the set top box. If the attackers capture the digital stream or record the video using camcorders, then the final content will include the information about the subscriber who received the content and ultimately is responsible for the theft of the digital assets. Once the content is recovered, and after a digital forensic reconstruction of the video stream, the data linking the content to the subscriber can be recovered and IPTV service providers can take the appropriate actions [15, 16].

## 6.5 Secure IPTV a Reality

Considering all the elements required for the operation of the IPTV service, it is easy to expect hundreds or even thousands of security vulnerabilities to be present when deploying the service for the first time.

Almost all applications used within the IPTV environment would be running on top of known operating systems. Only a few network components would have proprietary operating systems. The operating systems and applications would bring a significant number of security vulnerabilities, and it is unlikely that they would be patched by the vendor, requiring additional activities by the team working to secure the IPTV environment. All components must go through a process of validation before being deployed in the production environment.

In spite of the ingenuity of fraudsters and hackers and the vulnerabilities associated with a relatively new technology, secure IPTV is a viable and rapidly growing technology so long as security technology remains current as fraudsters implement their own improvements. The implementation of comprehensive security practices and processes can mitigate the risk involved and allow IPTV to take its place as a new and dynamic service to businesses and consumers.

IPTV must be protected using a point-to-point approach. Hackers will attempt to attack the infrastructure to gain access to contents or private information from subscribers using any vulnerable point.

Head-end elements have several vulnerabilities that are inherent to the underlying operating system used for the IPTV applications and communication equipment. The traffic within the head end must be restricted to specific VLANs, and all elements must be updated with the latest security patches.

IPTV allows users a great level of interaction with the IPTV applications. Some of those operate using HTTP and TFTP services. There are known vulnerabilities with those protocols and services, and hackers could exploit this to take control of servers and applications.

The home end is also vulnerable to hackers flooding the cable and sending attacks to home PCs and possibly to set top boxes. STBs have been modified to include hard drives and operating systems that allow greater interaction, but at the same time this increases the risk of intruders remotely controlling the appliances and capturing information.

From the content provider point of view, the STB must not be considered as a secure element. There are known cases of subscribers modifying the configuration of their modems, and it is likely that the same situation could happen with the STB.

Controls must be implemented to detect attempts to access contents, and actions should be taken to stop service to subscribers who have attempted to modify or have modified their systems.

All connections must be monitored to ensure subscribers are not rebroadcasting the contents received. In most cases the traffic will be sent using the same cable by which it was received, facilitating the detection process.

Subscribers will have to be trained to understand the new risks arising from using STBs with enhanced capabilities. This will allow them to recognize basic threats and will reduce the chances of unauthorized access to their STB systems. Subscribers must understand that they are responsible for their accounts and pin numbers, as well as being responsible for any content that is recorded or removed from the set top box.

A cost-effective IPTV security environment can be deployed on the basis of a detailed understanding of the technology involved and the communications between components. Security experts must consider the impact of removing or adding security mechanisms and must be aware of the existing threats to the environment.

# References

[1] National Institute of Standards and Technology. Available online: http://nvd.nist.gov/scap.cfm [2 October 2007].
[2] Mutz, M., 'Linux Encryption – How to?', 2007. Available online: http://encryptionhowto.sourceforge.net/Encryption-HOWTO-4.html [2 October 2007].
[3] Carnegie Mellon University, 'The Official CAPTCHA Site', 2000. Available online: http://www.captcha.net/ [2 October 2007].
[4] Ramachandran, K., 'Spoofed IGMP Report Denial of Service Vulnerability', 2002. Available online: http://www.cs.ucsb.edu/~krishna/igmp_dos/ [2 October 2007].
[5] Internet Engineering Task Force, 'An Overview of Source-Specific Multicast (SSM)', 2003. Available online: http://www.ietf.org/rfc/rfc3569.txt [2 October 2007].
[6] Cisco, 'MSDP MD5 Password Authentication'. Available online: http://www.cisco.com/en/US/products/ps6441/products_feature_guide09186a008049d556.html [2 October 2007].
[7] Internet Engineering Task Force, 'The Secure Real-time Transport Protocol (SRTP)', 2004. Available online: http://www.ietf.org/rfc/rfc3711.txt [2 October 2007].
[8] Internet Engineering Task Force, 'RSVP Cryptographic Authentication', 2000. Available online: http://tools.ietf.org/wg/rsvp/draft-ietf-rsvp-md5/rfc2747-from-07.diff.html [2 October 2007].
[9] Reeves, D., Nelson, B. and Clem, J. 'An Experiment Combining APOD and SE-SRVP'. Available online: http://apod.bbn.com/APOD-2_FTN%20presentation.ppt#482,5,RSVP Vulnerabilities [2 October 2007].
[10] US-CERT, 'Technical Cyber Security Alert TA07-005A', 2007. Available online: http://www.us-cert.gov/cas/techalerts/TA07-005A.html [2 October 2007].
[11] IEEE, '802.1X – Port Based Network Access Control', 1998. Available online: http://www.ieee802.org/1/pages/802.1x.html [2 October 2007].
[12] http://www.freedom-to-tinker.com/?p=1005! [2 October 2007].
[13] Crosby, S. and Goldberg, I., 'A Cryptanalysis of the High-bandwidth Digital Content Protection System'. Available online: http://citeseer.ist.psu.edu/cache/papers/cs/25911/http:zSzzSzwww.cs.berkeley.eduzSz~dawzSzpaperszSzhdcp-drm01.pdf/a-cryptanalysis-of-the.pdf [2 October 2007].

[14] 'High Bandwidth Digital Content Protection System', 2006. Available online: http://www.digital-cp.com/home/HDCP_Specification%20Rev1_3.pdf [2 October 2007].

[15] '5C Digital Transmission Content Protection', 1998. Available online: http://www.dtcp.com/data/wp_spec.pdf [2 October 2007].

[16] Chiang, A., 'Integrating System-wide Security into SOC Designs', 2007. Available online: http://www.elecdesign.com/Articles/Index.cfm?AD = 1&ArticleID = 15383 [2 October 2007].

# Appendix 1

## Converged Video Security

### A1.1 Introduction

Although IPTV has created new service and revenue opportunities for traditional voice and data carriers, the technology also comes with a high level of risk owing to attacks by fraudsters and hackers. However, with the proper planning, development and implementation of effective security processes, this risk can be managed efficiently and cost effectively.

### A1.2 Threats to IPTV Deployments

Compared with traditional voice/data networks or cable TV infrastructures, threats to an IPTV environment are far more severe. IPTV allows carriers to manage valuable content that must be protected from unauthorized access and modification. Carriers also need to ensure that quality of service is protected to comply with customer's expectations and service level agreements (SLAs).

For years the satellite TV industry has been fighting access fraud.[1] Recently, satellite TV companies have been taking legal action against defendants for unauthorized access to TV content.[2]

The experience of the satellite TV industry shows that fraudsters go to great lengths to break their security measures. This includes cracking the smart card protection used for the set top boxes and distributing cloned 'free access' cards. Even though the satellite TV providers have modified the cards, fraudsters have managed to find alternative ways to break the safeguards incorporated in the new releases.

Now that video technology has entered the IP world, the level of threats has escalated – vulnerabilities that have been solved in other, more mature technologies are still part of the new IPTV systems.

Figure A1.1 shows a standard infrastructure for an IPTV network environment.

---

[1]  See http://www.theregister.co.uk/2004/12/11/directv_hacker_sentenced/
[2]  See http://www.directv.com/DTVAPP/aboutus/headline.dsp?id=03_03_2005A

---

*IPTV Security: Protecting High-Value Digital Contents*   David Ramirez
© 2008 Alcatel-Lucent. All Rights Reserved

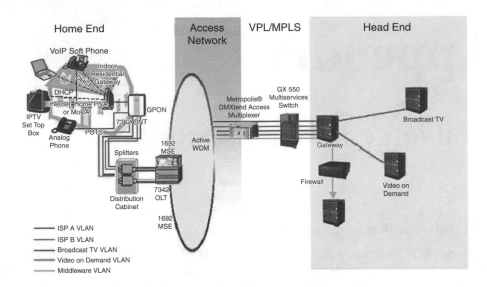

**Figure A1.1** IPTV network architecture

IPTV is transferred not only to set top boxes but also to computers and handheld devices, which facilitates hacker access. Simple software modifications introduced by hackers allow them to break the encryption system and other security measures, or even capture and redistribute the contents using peer-to-peer networks.

A major impact on the satellite TV industry has been fraudsters selling modified 'all access' smart cards. As a result, the IPTV industry faces an entirely new threat – with broadcasting stations residing on every home PC, hackers are able to redistribute the broadcast stream to other computers all over the world.

## A1.3 Protecting Intellectual Property

There are currently three primary types of technology used to protect video application intellectual property (IP) rights: content protection systems, conditional access systems and digital rights management.

*(i) Content Protection Systems (CPSs)*
Content is transmitted across networks in an encrypted form to help protect against theft or unauthorized access.

Content protection systems are used to help ensure that content is only viewed by authorized subscribers. Even if an intruder has access to the communication, content should be encrypted. Handling of the key management process, including how keys are exchanged between parties and change frequency, is an important security issue.

*(ii) Conditional Access Systems (CASs)*

A CAS helps ensure that only authorized subscribers have access to the content, creating a safeguard against theft of service.

These systems are used by cable operators to control access to content. CAS has evolved from simple frequency shifts and electronic noise to content encryption. Because telcos and carriers face different threats associated with IPTV, a variety of new technologies have been developed to protect data.

*(iii) Digital Rights Management (DRM)*

Content owners realize that IPTV provides a great channel but is also a huge risk. The growth of peer-to-peer networks demonstrates that digital content can easily be traded on the Internet with little content owner control and without proper recognition of IP.

DRM manages how the content is used by the subscriber on the basis of specific conditions set by the distribution contract.

Technology providers initially implemented solutions that lacked strong DRM control. Today, in spite of current baseline DRM controls, some vendors are still failing to address the issue of recording and replay – subscribers can store copies of DRM material and redistribute it on the web. This has prompted content owners to take matters into their own hands and begin implementing strong DRM controls to reduce the risk of unauthorized access to content.

## A1.4  VOD and Broadcast

There are currently two basic options for content delivery using IPTV: video on demand (VOD) and broadcast. Each has its own particular security and DRM requirements.

For VOD it is generally recommended that the content be segmented and encrypted using a symmetric key. The key can be changed several times in a typical movie to increase protection. Each set top box has the subscriber's private key, and the VOD server sends the encrypted content and the encrypted symmetric keys to the set top box for decryption and playback.

Broadcast content follows a very similar process. Content is encrypted at the source with a symmetric key. The set top box sends a request for the current content key, and the content server sends an encrypted symmetric key to be used by the set top box to retrieve the contents.

Current industry trends indicate that content owners are demanding relatively strong encryption for their data, including the implementation of the advanced encryption standard (AES).[3] This will probably shift attacks from content to the endpoints and the transport layer.

Other DRM requirements include the stipulation that DRM-protected information should be unintelligible after leaving the source. It should only be decrypted after it has arrived at the destination. This involves changing the security architecture to include a key repository that is used as part of the encryption process.

The new protection technologies, included also on high-definition television (HDTV), bring a series of standards that ensure digital transmission and protection from the sources to the TV. This allows additional protection against unauthorized copies at the subscriber side.

---

[3] See http://www.screenplays.bz/sp105o.html

## A1.5 Smart Cards and DRM

Although other technologies are also in use, smart cards are one of the most widely used technologies to facilitate the authentication process and protect video over IP. These cards contain an embedded microprocessor that can be used to store security information such as private keys for digital signatures. Without a smart card there is no second-level authentication. This allows hackers to simulate the details from a valid subscriber and steal access to the service.

The satellite and cable TV industries have used smart cards for some time and, even with known cases of card fraud, recognize that the use of cards is one of the best ways to reduce the risks of unauthorized access and content fraud.

DRM systems are currently being integrated with smart cards and private key storage to enable the encryption of content from the source and allow the creation of specific streams for each subscriber. Smart cards are used to authenticate the subscriber, which allows service operators to encrypt packets and send the key to the subscriber's set top box.

The smart card also facilitates the process of using different encryption keys during the broadcast process, thus increasing the complexity and resources required to break the security of the content.

Even after the smart card encryption has been breached, fraudsters have the added problem of distributing the cloned cards. The logistics related to this process create a barrier and reduce the impact of access fraud. (Without the cards, fraudsters can simply distribute software modules that enable unauthorized access to the contents. A similar situation occurs with regard to the 'content scramble system' (CSS) protection implemented for DVDs.[4])

DRM has also been implemented on iTunes, and there are claims that a free piece of software[5] can be used to access DRM-protected files and create an unprotected version. Similar situations have affected other DRM vendors, allowing subscribers to remove the DRM protections of music and video files.

## A1.6 Countering the Threats

A very effective way to analyze the threats to the infrastructure is to use the ISO/IEC 18028-ITU X.805[6,7] framework to develop a comprehensive threat model for a generic converged video infrastructure.

Some of the threats to this type of network include:

---

[4] For example, Jon Johansen, a Norwegian, posted software on the Internet that allowed anyone to open the CCS protection and access the DVD's contents.
[5] See http://www.videolan.org/
[6] See http://www.itu.int/home/index.html
[7] *ITU X.805 press release* (http://www.lucent.com/press/0304/040317.bla.html):

> 'It is paramount that security be a well-thought process that goes from system inception and design to system implementation to policies and practices for system deployment', said Houlin Zhao, director of the Telecommunication Standardization Bureau for the ITU. 'Lucent and Bell Labs understand security, and they recognize the importance of a standards-based approach for realizing it. I applaud their work in driving the development and adoption of Recommendation X.805 – a significant step on the path to securing networks worldwide'.

- denial of service attacks and worm propagation;
- network infrastructure attacks;
- Trojan horse programs and customer's theft of service;
- self-provisioning infrastructure attacks;
- billing infrastructure attacks;
- intellectual property (IP) theft.

Lucent Bell Labs played an important role in the development of ITU X.805. The ITU X.805 recommendation provides a framework for a thorough review of all aspects of security in an IPTV solution.

There are eight security dimensions that cut across the management, control and end-user security planes of the X.805 framework. Each plane has its own underlying infrastructure, services and applications. This encompasses everything from the IP backbone through the video application middleware, to the video head office and finally the set top box.

The eight dimensions are:

- access control;
- authentication;
- nonrepudiation;
- data confidentiality;
- communication security;
- data integrity;
- availability;
- privacy.

The intersection of each security layer with each security plane indicates where security dimensions are applied to counteract the threats.

Figure A1.2 graphically shows the security layers, planes, dimensions and threats as defined by the ITU-TX.805.

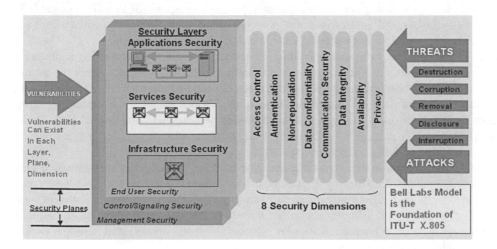

**Figure A1.2**  Applying security dimensions to security layers

**Table A1.1**  Mapping of security dimensions to security threats

| Security dimension | Security threat | | | | |
|---|---|---|---|---|---|
| | Destruction of information or other resources | Corruption or modification of information | Theft, removal or loss of information and other resources | Disclosure of information | Interruption of services |
| Access control | Y | Y | Y | Y | |
| Authentication | | | Y | Y | |
| Nonrepudiation | Y | Y | Y | Y | Y |
| Data confidentiality | | | Y | Y | |
| Communication security | | | Y | Y | |
| Data integrity | Y | Y | | | |
| Availability | Y | | | | Y |
| Privacy | | | | Y | |

Table A1.1 provides a mapping of security dimensions to security threats. The mapping is the same for each security perspective. The letter 'Y' indicates that a particular threat is countered by a corresponding security dimension.

There are several threats to the IPTV service, but, in comparison with other broadcast technologies, the security implementation is relatively easy.

Some of the threats that should be included as part of an IPTV risk assessment include the following.

## A1.6.1 Threat References

When unscrupulous advertisers started sending unsolicited emails to thousands of email addresses, it was called SPAM. Then instant message systems were targeted, and that was termed SPIM (SPAM over IM). Voice-over IP can also be a target for unsolicited calls, and it too has its own name – SPIT (SPIT Spam over IP Telephony).

Now, IPTV technology needs to implement protections against SPIV (SPAM over IPTV). If set top boxes and applications are not configured to authenticate/validate sources of content, they will end up displaying unsolicited pop-up advertisements.

IPTV is supported by known operating systems and commercial networking equipment. In the case of viruses and worms, a disruption can be caused to the service either by saturation of the networks or by crashing the network elements and endpoints (STBs). To prevent these incidents, normal patching and testing is required. Worms might crash supporting services such as billing or provisioning, middleware servers or head-end components. If the incident happens during a particular VOD selling time, then the vendor will suffer a revenue loss, not being able to fulfill subscriber requests.

As content becomes more valuable and entertainment systems begin to integrate known operating systems, fraudsters will attempt to use Trojan horses to steal access to content. Subscribers might inadvertently install software that allows intruders to gain access to content and even request VOD using the subscriber's account. It is difficult to avoid this situation, and it requires a constant interaction with the system, automatic updates and antivirus packages to maintain systems protected.

Intruders might try to control the provisioning infrastructure or the billing infrastructure as part of a fraud attempt either by creating 'ghost' accounts or by changing the entries on the billing system. If proper auditing and monitoring systems are not implemented, intruders might be able to change information from the central applications.

## A1.6.2 Threat Models

The following threat models describe the different layers and planes of a video IP infrastructure.

*(i) Application Layer Management Plane Threat Model*
Securing the management plane of the application layer is accomplished by controlling the application management data.

As part of securing the management plane, the following controls should be implemented when securing the application layer management plane:

- *Access control.* Ensure that only authorized personnel are allowed to perform administrative activities on the application, such as administering VOD requests and having access to content lists and descriptions. Only authorized users must undertake the remote management of set top boxes, central office equipment, broadcast network equipment and different elements related to the broadcast head end. Intruders planning to manipulate the configuration of the different elements can use default parameters and vulnerable services. Lack of appropriate access controls could result in DOS attacks, access fraud, and man-in-the-middle attacks.
- *Authentication.* Verify the identity of the person or device attempting to perform network application administrative activities. Hackers will target the applications in attempts to gain unauthorized access to contents and data. Identities must be verified to minimize the risk of unauthorized access to contents. Access control is linked with the authentication of users. The identity of all system users must be confirmed.
- *Nonrepudiation.* Provide a record of all individuals performing administrative actions on network applications. This control also helps reduce the incidence of subscriber fraud; for example, service provisioning to 'ghost' accounts. Most system updates are carried out using automated tools – set top boxes and networking elements must require nonrepudiation parameters from the management terminal. This type of transaction could involve digital signatures of configuration scripts, or it could involve the use of management terminals to establish a secure connection before sending the information.
- *Data confidentiality.* Protect all files used in the creation and management of the application. Controls to reduce the risks of information disclosure should be implemented in order to protect customer details and administrative information. All customer information must be protected from third-party access, including electronic records of customers and configuration information which should be encrypted to avoid unauthorized access.
- *Communication security.* Ensure that the management information flows only between the application and an authorized party. Because worms and hackers may attempt to send management information to different network elements, controls should be implemented to

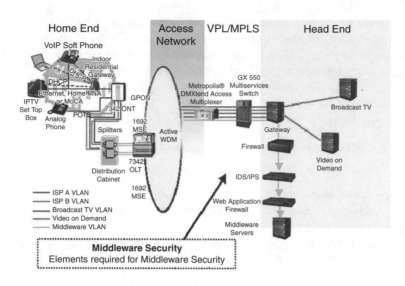

**Figure A1.3**  ITU-T X.805 sample threat model for video IP

avoid third parties gaining access to management communications. Figure A1.3 shows how the middleware should be protected by deploying a number of layered counter measures. Man-in-the-middle attacks can potentially be prevented by protecting communication security. Intruders could attempt to capture information sent by the management terminal to the various network elements. Set top boxes send and receive configuration information that can be manipulated by intruders if not properly secured.

- *Data integrity*. All configuration and management information and data from the application should be protected from unauthorized modification. This prevents intruders from modifying the data and causing system failures.
- *Availability*. Ensure that the ability to administer the application is not affected or denied.
- *Privacy*. Ensure that information that can be used to identify network-based applications is not disclosed to unauthorized parties.

*(ii) Application Layer Control Plane Threat Model*
Securing the control of signaling information supports securing the control plane of the application layer.

The following controls should be implemented when securing the application layer control plane:

- *Access control*. Ensure that application control information received by a network device participating in a network application originates from an authorized source. The different parameters received should be protected, as third parties could manipulate part of the control process. It is very important to implement access controls so only valid elements are used for particular services such as DHCP, DNS and others.
- *Authentication*. Verify the identity of the origination of application control information sent to network devices participating in the application.

- *Nonrepudiation.* Provide a record identifying the person originating the application control information. All control information should include nonrepudiation data. This involves either digital signatures or log details of the system and account used to deliver the information.
- *Data confidentiality.* Protect application control information resident in a network device being transported across the network. Application data should be protected from third-party access, as intruders could try to capture control information, session keys and other data.
- *Communication security.* Ensure that application control information being transported only flows between authorized nodes. This includes protection against man-in-the-middle attacks.

*(iii) Application Layer End-user Plane Threat Model*
Securing the end-user plane of the application layer includes securing user data provided to the network-based application.

The following controls should be implemented when securing the application layer end-user plane:

- *Access control.* Ensure that only authorized users are allowed to use the network application.
- *Authentication.* Verify the identity of the user attempting to use the application.
- *Nonrepudiation.* Provide a record identifying each user accessing the application.
- *Data confidentiality.* Protect end-user data that is being transported or stored by the network-based application.
- *Communication security.* Ensure that end-user data is not diverted or captured by unauthorized third parties.
- *Data integrity.* Protect end-user data that is being transported by a network-based application against modification, deletion or replication.
- *Availability.* Ensure that access by authorized end-users to a network-based application is not denied.
- *Privacy.* Ensure that the network-based application does not disclose information about the end-user.

*(iv) Services Layer, Management Plane Threat Model*
Securing the management plane of the services layer includes securing the operation, administration, maintenance and provisioning of network services.

The following controls should be implemented when securing the service layer management plane:

- *Access control.* Ensure that only authorized personnel and devices are allowed to perform network service management activities – for example, granting a user access to the service.
- *Authentication.* A person's identity should be verified before access is allowed to administrative functions on the service layer.
- *Nonrepudiation.* A record should be kept of all individuals performing network services administrative tasks.
- *Data confidentiality.* The network services management information should be protected from unauthorized access – this includes passwords, configuration and parameters.

- *Data integrity*. Management information of network services should be protected against unauthorized modification, deletion or replication.
- *Availability*. Measures should be implemented to ensure the ability of administrators to manage the network services at all times.
- *Privacy*. Ensure that information that can identify the administrative management systems is not available to unauthorized personnel.

*(v) Services Layer Control Plane Threat Model*
Securing the control plane of the services layer includes securing the control of signaling information used by the network services.

The following controls should be implemented when securing the services layer control plane:

- *Access control*. Ensure that all control information received by the different elements has been sent by an authorized source.
- *Authentication*. The identity of any element sending control information should be verified.
- *Nonrepudiation*. A record should be kept, identifying all elements sending control information.
- *Data confidentiality*. All control information stored or sent across the network should be protected against unauthorized access.
- *Communication security*. All service control information should flow only between the intended source and destination.
- *Data integrity*. Service control information held in network devices or servers should be protected against unauthorized modification.

*(vi) Infrastructure Layer Management Plane Threat Model*
Securing the management plane of the infrastructure layer includes securing the operations, administration, maintenance and provisioning of the individual network. The configuration of an individual switch is a typical management plane activity.

The following controls should be implemented when addressing the security of the infrastructure layer management plane:

- *Access control*. Ensure that only authorized personnel or devices are allowed to perform administrative activities on the network devices or communications link. This control applies to all the various elements involved in the communication. Without proper access controls, intruders could modify the configuration of communication elements and cause service interruptions, steal service, or create a man-in-the-middle attack.
- *Authentication*. Verify the identity of the person or device performing the management activities on the network elements. Proper authentication of devices helps reduce the impact of man-in the-middle attacks and ensures that downstream elements accept information only from authenticated elements. Worms tend to rely on unauthenticated communications to infect other elements.
- *Nonrepudiation*. Maintain a record of all individuals or devices that undertake administrative activities on the network elements. Any change to the system must be matched against an individual or device. This improves auditing capabilities and facilitates any forensic investigation.

- *Data confidentiality.* Protect network communications against unauthorized access or viewing. This involves all data related to the infrastructure layer, including configuration information, authentication data and backup data. If configuration information is sent without protection, intruders may be able to capture configuration and authentication information.
- *Communication security.* Data flow should be protected for secure remote management and administration. Ensure that communications are not diverted or intercepted. This control is also important when protecting against man-in-the-middle attacks.
- *Data integrity.* Administrative and management information must be protected against modification to prevent intruders from modifying instructions and configuration data. Measures should be implemented to avoid changes to configuration data.
- *Availability.* The ability to manage the network devices or communications link should be protected. Information about the configuration of the different systems should be available at all times. High availability and distributed infrastructures should be implemented to reduce the impact of DOS attacks.
- *Privacy.* Information that can be used to identify network devices should not be available to unauthorized parties. This reduces the risk of intruders being able to map the network.

*(vii) Infrastructure Layer Control Plane Threat Model*
This plane is concerned with securing the signaling information that resides on the network elements and server platforms.

The following controls should be addressed when analyzing the infrastructure layer control plane security:

- *Access control.* Network devices should only accept control information messages from authorized network devices – this reduces the threat of unwanted modifications. It also helps prevent worms from using control information to reconfigure the infrastructure and cause a denial of service.
- *Authentication.* The identity of the device sending control information should be confirmed. This control works in conjunction with access control to protect against unwanted changes.
- *Nonrepudiation.* Maintain records that identify devices sending control information. Any modification should be recorded with the identity of the issuer, ensuring accountability for any changes to the infrastructure.
- *Data confidentiality.* Control information includes data that is considered confidential, such as passwords and security configuration details. To avoid third-party access to control information, data should be protected.
- *Communication security.* Ensure that control information flows only from the intended source to the specified destination. Intruders may try to stop or capture signaling information to cause a denial of service.
- *Data integrity.* Protect control information stored in network devices in transit or held at the servers. Intruders may attempt to capture control information in order to reconfigure the network elements.
- *Availability.* Ensure that network elements are always able to receive control information. Intruders may try to cause a denial of service by flooding the authentication server or other critical systems.
- *Privacy.* Information that could be used to identify a specific network element should be kept confidential.

# Appendix 2

## Federated Identity in IPTV Environments

### A2.1 Introduction

As broadband access becomes commonplace, IPTV will experience rapid growth. However, a flexible, scalable methodology for subscriber authentication is essential. Two federated identity solutions now available from the Liberty Alliance Project and the open-source Shibboleth community – both using the SAML (Security Assertion Markup Language) protocol – provide a foundation for deploying secure, user-friendly authentication services.

### A2.2 IPTV Federated Identity Solutions

Owing to bandwidth limitations during the first half of this decade, video content was slow to make its impact felt on the web. Now, however, with the ubiquity of DSL and cable connections, plus new mobile data services such as WiMAX, 3G and UMTS, the general public is quickly adopting this previously inaccessible new media.

IPTV presents service providers with the opportunity to embed a variety of services within a single medium. But this opportunity is not without its challenges. Critical to deploying IPTV is the authentication of subscribers using a significant number of different applications and systems. In most cases the authentication will include third parties. These involve specific content owners, such as providers of ring tones, movies, short clips and shared access to content, that are working with the service provider or even independently.

A standards-based resolution of these identity and authentication problems is addressed by federated identity solutions. Currently, two approaches have taken the lead – one from the Liberty Alliance, and the other from the open-source Shibboleth community.

Based on our previous experiences and understanding of the building blocks for a successful federated identity solution, we feel it is important carefully to review the specific federated identity management model before implementation. This includes mapping the existing elements against the requirements for a federated identity architecture. It also means creating a blueprint indicating the areas already covered by existing identity management (IdM) solutions and those areas that must be implemented. Most organizations already have many of the elements required for deploying a federated identity solution; they just need to articulate the elements and add the missing pieces.

Federation standards play a critical role in enabling cross-departmental, cross-organizational and cross-security domain integration. By leveraging an existing standard model, cross-domain identity integration can be attained as part of a cost-effective and repeatable process.

## A2.2.1 SAML for Security

At the core of the federated identity model is SAML (Security Assertion Markup Language). Developed by the Security Services Technical Committee of OASIS, SAML is an XML-based framework for communicating user authentication, entitlement and attribute information. SAML allows business entities to make assertions regarding the identity, attributes and entitlements of a subject.

### A2.2.1.1 The SAML Process

When two entities need to exchange information about a particular individual, they can use SAML. Entity A issues the SAML assertion. Subsequently, the information is received, verified and authorized by entity B.

Figure A2.1 shows how the two parties exchange SAML traffic to validate and authorize the access of an individual from entity A.

The process starts when the individual from entity A requests access to a local resource ①. The request is validated against the authentication and identity store ②. This store is usually an LDAP-compatible system that stores all the identities valid in the security domain, and can validate the identity of individuals. The authentication and identity store confirms to the resource (web server or other resource) that the individual holds a valid identity in the security domain ③. The resource then proceeds to issue a valid token. The individual holding a valid token within entity A selects a resource outside the security domain and is redirected to the federation server ④. Here the token is used to create the SAML assertion and also validate the information against the authentication and identity store.

The individual is redirected to the remote federation server ⑤. This traffic includes the SAML assertion and is received by the federation server operated by entity B. The information received is validated against the identity and authentication stores ⑥. Rather than retrieving information from that particular individual, information that is not held locally, the identity store is used to retrieve, map or provision user identifiers and attributes based on the information received within the SAML assertion ⑦.

Finally, the individual is redirected to the resource requested, along with a session token that has been validated and authorized. The token contains the identifiers and attributes previously agreed with entity A for that particular type of user ⑧.

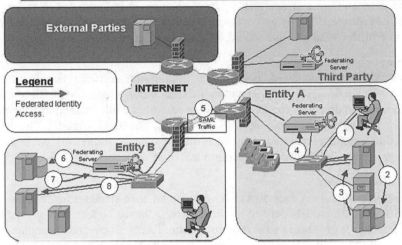

**Figure A2.1**   Federated identity – SAML traffic

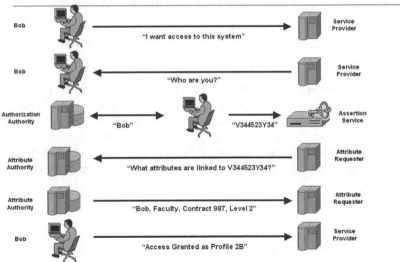

**Figure A2.2**   Shibboleth SAML flow

Figure A2.2 shows the shibboleth SAML flow required for authenticating and authorizing a user.

Many of the required elements already exist as part of an IdM operated by most entities. It is important to establish which additional elements are required and then document descriptions of the requirements in order to implement the federated identity architecture.

### A2.2.1.2 Reviewing Existing Standards

To ensure compatibility and operability of the solution, it is important to review existing standards for federated identity and compare those against the existing IdM operated by the entity. Some of the standards that should be considered are as follows.

*Shibboleth*
Shibboleth is a standards-based, open-source middleware software that can be used to implement web single sign-on (SSO) within different security domains. It allows entities to make authorization decisions for access to online resources while maintaining the privacy of individuals. Shibboleth is compatible with the OASIS SAML v1.1 model. This technology can be used to support a federated SSO and attribute exchange framework while maintaining control over the attribute information being released to each party involved.

*Liberty Alliance Project*
This industry consortium is dedicated to developing an open standard for federated network identity. Some of the models defined by the Liberty Alliance concentrate on enabling SSO through the concept of identity federation or account linkage. Once two accounts have been linked, an individual is able to use one account after authenticating to the other. The protocols and messages that enable this SSO between the first and second sites are based on the SAML protocol.

## A2.3 Applicability to an IPTV Security Environment

IPTV environments can be used to provide subscribers with access to a significant number of services from different sources. There are two main applications that benefit from federated identity solutions:

- **internal applications within the IPTV service provider** – subscribers requiring access to a variety of internal applications from the service provider, including help desk, customer management and retention, account information and access to general services such as email and Internet browsing;
- **external applications provided by partners** – subscribers wanting to have access to third-party content, including music videos, movies, local services, home delivery, and purchases, as well as government services.

### A2.3.1 Internal Applications

These applications include subscribers needing access to a significant number of applications from different security domains. Requesting authentication credentials each time a subscriber needs to move from one domain to another adds complexity and time to the process and negatively affects the customer's experience.

Some basic domains include:

- help desk site (managed by the operations team);
- customer retention site (managed by the marketing or commercial team);
- account information site (managed by the accounts department);
- email and Internet sites (managed by the IT department).

Each security domain stores sensitive information that must be segregated and controlled to reduce the risk of abuse or fraud and maintain compliance with privacy laws. The simplest way to maintain segregated control over access and privileges is deploying a federated identity infrastructure.

With a federated identity model, each security domain has control over the profiles enabled within the system and the levels of access granted to each profile. A security policy can be defined, authorizing specific access to normal subscribers and additional levels granted to administrators, help desk and operators. This approach guarantees that each individual will have access based on a clearly defined model, reducing the possibilities of unauthorized access by accident, errors or security break-ins. Domains and profiles can even cover regions, zones or cities, with a particular geographical region being allowed to access only records that correspond to that region. When a user attempts to access records from other areas, the system locks the request.

All web based applications within the service provider can be enabled to support the federated identity validation and authentication. With this additional code added to the web applications, access is granted only to users that fit the approved profile.

## A2.3.2 Set Top Box Security

The nature of IPTV implies that most subscribers will use their television set to browse information and applications (other compatible access models such as PCs and handheld devices will also be used, but to a lesser degree.) Access is based on the browser provided by the set top box (STB). Cryptographic functions and hardware within the STB can be used to store credentials or at least partial authentication information. This approach reduces the amount of information requested of subscribers.

From the security domain point of view, the servers receive an access request from the subscriber STB. A set of stored credentials are matched against the authorization authority that issues the subscriber profile information. This information is sent by the assertion service to the attribute requested which, in turn, validates the information against the attribute authority. The attribute authority responds to the attribute requester with a set of entries from the approved schema. This information is used to grant access based on the approved profile model.

A subscriber will be able to use the same credentials against a number of servers configured to support the federated identity model. For example, a subscriber browsing the help desk site will be able to jump to the accounts application by selecting an option from the online menu. The subscriber can also access emails and request third-party services, invoicing and validation relying on the federated identity model.

## A2.4 Video on Demand

The rapid growth of peer-to-peer services in the past 10 years indicates that users want flexible access to content. High-quality and easy-to-use delivery mechanisms can generate significant revenues. The normal evolution of the technology will bring a demand for video content any time, anywhere over any device – this includes portable devices receiving broadcasts over wireless high-speed networks. Users will also demand access to all available content, not just the normal sequential channels with advertising.

Video on demand will be complemented with channels on demand. Users will have the flexibility of subscribing to channels not provided as part of their main content package. This brings new players into the market – content aggregators and brokers will offer a global content in exchange for a fee that will be shared with the content owners.

This new environment made up of users requesting all type of content when and where they want it requires a very flexible identity management architecture. Users will not accept a service that asks them to present a user name and password each time they want to change the channel. The current state-of-the-art way to meet this requirement is the use of a federated identity model that facilitates a cross-authentication between environments and domains, a model that is flexible enough to be commercially acceptable by non-technical users.

# Appendix 3

## Barbarians at the Gate

### A3.1 Barbarians at the Gate

During the Roman Empire, the expression 'barbarians at the gate' referred to an imminent attack by invading armies. Today, service providers addressing the issue of IPTV security have their own barbarians to deal with, and the digital supplier line access multiplexer (DSLAM) became the border outpost located between the service provider's known and trusted environment and the external space, which is unreliable and potentially hostile.

Most subscribers are trustworthy citizens who duly pay for any of the services they receive – they never consider manipulating their equipment's configuration for financial gain. However, there are always a small number of potential intruders who are at the gates trying to break through and gain unauthorized access to content. These latter-day barbarians are the reason why service providers must concentrate on implementing appropriate security mechanisms that eliminate or reduce the cost and other unpleasant consequences of hacker attacks.

The Roman Empire lasted a little over 500 years. It finally collapsed in 476 CE from a series of chain reactions triggered by events that happened decades, even hundreds of years, earlier. Today's timescales are far more compressed. When an IPTV security incident occurs, service providers can be certain that in a mater of seconds a poorly secured environment will be either taken down or commandeered by intruders.

Consider the following scenario. A pirate TV station has broken your security and is using your network, equipment and signal to broadcast antigovernment or pornographic content. How much will it cost to correct the situation? How will your subscribers respond? What kind of regulatory backlash will the incident trigger? And whose jobs are going to be on the line? These are just a few of the questions that senior management can anticipate having to deal with, especially if they decide to postpone implementing adequate security measures until 'phase II' (phase II is that indeterminate state that occurs after the people who need to do the work say something like 'I'm too busy', or 'we don't have the budget' or even more shortsightedly 'Nothing has happened before, so what's the problem?'.

*IPTV Security: Protecting High-Value Digital Contents*   David Ramirez
© 2008 Alcatel-Lucent. All Rights Reserved

**IPTV Security Architecture**

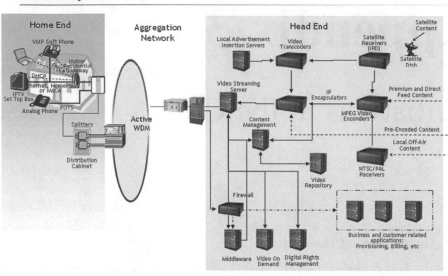

Figure A3.1   IPTV infrastructure showing areas where intruders can concentrate their attacks

Figure A3.1 shows the detailed view of the IPTV infrastructure, including critical elements for the operation of the service.

## A3.2  How to Break an IPTV Environment

An undersecured IPTV environment is an open invitation to the hacker. There are access elements that can be manipulated, a transport network that demands very high bandwidth and middleware with web technologies that can be broken into in a matter of seconds. Within the head-end environment there are the crown jewels – thousands and thousands of DVD-quality movies at the fingertips of a hacker or pirate wanting to make a few bucks.

## A3.3  Network Under Siege

Attacks on the network can compromise confidentiality, integrity and availability.

### A3.3.1  Confidentiality

There are three basic levels of content that must be kept confidential:

- *Customer-related information.* This includes account information, contact details, credit card numbers and past purchases of video on demand (VOD) – people prefer to keep their taste in movies a private matter.
- *Customer DRM information.* This includes the keys and temporary access tokens required by the subscribers to access DRM-protected content. Third parties may be able to access

content using these keys. Even worse, hackers may be able to broadcast the content without the DRM protections; for example, broadcasting concerts, sports events and other pay-per-view (PPV) content for free over the Internet.

- *Content information*. Content owners trust service providers and expect that their valuable content will be protected. Their business model relies on the protection of valuable assets and limiting access to these assets to paying customers only.

## A3.3.2 Integrity

- *Customer-related information*. Intruders may be able to modify records or customer information, e.g. by adding or removing payments to the account or services provided. A common fraud gambit by unscrupulous third parties is an offer to 'settle' a customer account for a payment equivalent to just a percentage of the money owed to the service provider. Or, a hacker with access to or control over a particular broadcast may manipulate the system to request a particular TV package on behalf of his next-door neighbor, capture the broadcast in transit and leave the bill for his neighbor or someone else.
- *Content information*. A simple example is the potential modification of content. For example, hackers may manipulate the source content and insert immoral, political or even commercial information into a PPV Super Bowl broadcast or a concert by Sting. The result is SPIT (Spam over IPTV), with hackers adding promotions for, say, Viagra® or the hottest new penny stocks.

## A3.3.3 Availability

- *Customer-related information*. Customers need access to their self-service interface to request packages and content, or to review and settle their accounts. An attack on availability may take down the middleware server in charge of self-service and leave the broadcast and VOD servers running. Any impact on the customer experience is negative and may affect revenue from and customer confidence in what is basically a new type of service.
- *Customer DRM information*. If customers are denied access to DRM keys, they have no access to content. Customers will blame their service provider because, while they know that they are connected, the set top box is telling them that the desired content cannot be accessed and opened.
- *Content information*. There are many other scenarios that could deny the subscriber access to content. Just a few include: hackers causing a denial of service (DOS) on the access/aggregation network; hackers shutting down or causing a DOS on the middleware; and hackers removing or destroying the contents.

## A3.4 Countermeasures

Many different scenarios can occur if the security of the IPTV environment is not appropriate. Two important security layers are the set top box (STB) and DSLAM.

### A3.4.1 Set Top Box

For example, within a layered security environment, the first layer to be considered is the set top box (STB). This element actually stores the DRM client and keys, and is the point of interaction with the subscriber. Although this is a critical element that requires the implementation of security measures, because the STB is within the customer premises, it is beyond the control of the service provider.

## A3.4.2 DSLAM

The second security layer, the DSLAM, is held within a protected physical environment where the logical configuration can be protected and ensured. This is the service provider's primary line of defense.

There are three main functions within a DSLAM:

- *IP concentration*. This function covers the concentration of data received by the other functions and the preparation of data for transmission.
- *IP services*. This function covers the advanced IP functionality and includes areas such as routing and bridging. Layer-2 and layer-3 capabilities are found within this function. Most of the security mechanisms available within the DSLAM are found here.
- *DSL line service*. This function covers the physical link with the customer's premise termination equipment.

Most of the DSLAM security characteristics reside in the IP services function. The key aspects to consider are:

- access and session control;
- routing;
- user segregation;
- quality of service (QoS);
- virtual networks and virtual service.

### Access and Session Control

The DSLAM supports a series of subscriber authentication and session tracking capabilities. Subscribers are authenticated against a local user database or through an external RADIUS server. Subscriber and physical link data can be considered during the authentication process – only authenticated physical ports are allowed to exchange information with the transport network and head end.

DSLAMs can validate and authorize users on the basis of the MAC address used on the other side of the physical link, as well as other common authentication mechanisms such as challenge handshake authentication protocol, IP address authentication and PPP authentication protocol.

In cases where DHCP is used to assign IP addresses, the DSLAM can use DHCP option 82 to insert data related to the physical link. The authentication is primarily based on the DHCP option 82 protocol. DHCP option 82 snoops the DHCP request coming from the subscriber network and inserts the physical line identification information in the option 82 field of the DHCP package. This information typically includes access node ID, shelf ID, slot ID, and finally the line ID. Processed more deeply in the network by the DHCP server, this option is used to identify the subscriber and assign (or refuse to assign) a valid IP address for the subscribed services.

**Table A3.1** How ACL security ensures confidentiality, integrity and availability

|  | Confidentiality | Integrity | Availability |
| --- | --- | --- | --- |
| Access and session control. | DHCP option 82 reduces the chance of MAC and IP spoofing. | Attackers are not able to spoof a valid address in order to attack the head end and modify information. | Attackers are not able to spoof a valid address in order to launch DOS attacks. |

Table A3.1 provides references to how ACLs improve security.

If a subscriber attempts to bypass the authentication provided by the DSLAM, the system easily recognizes that the subscriber has manipulated parameters. For example, suppose user A attempts to spoof a valid MAC address from subscriber B by modifying his own MAC address and replacing it with the one from B. In this case, the DHCP server will identify that the MAC address submitted by subscriber A during the DHCP initial request corresponds to a MAC address already registered with a different physical location (the physical location of B).

DHCP option 82 is a powerful mechanism that can be used to ensure tighter control over access to the infrastructure. The approved MAC addresses can be stored with the physical information, allowing each subscriber to have a small number of network elements preauthorized to access the network. Each time a new MAC (network element) attempts to join the network, the system may ask for authentication information (a user name and password) before adding that particular MAC to the database. Properly deployed, DHCP option 82 will eliminate most opportunities for MAC and IP spoofing. It will also increase the level of confidence with regard to system information, and also reduces the incidence, frequency and impact of unauthorized access to content.

DSLAMs can also provide access control lists (ACLs) and usage restrictions to control the type and amount of data sent by subscribers.

## A3.4.3 Routing

Service providers can use layer-3 static, dynamic, and policy-based routing. Policy-based routing is implemented using several virtual routing domains (VRDs). Each VRD is used to provide the routing decision for subscriber packets and is based on the routing entries of the VRD that the subscriber belongs to.

With the basic routing, packets can be segmented within valid domains only, ensuring that only authorized network elements transmit within the IPTV VRD.

Routing can be used to provide mechanisms that protect confidentiality, integrity and availability.

Table A3.2 provides details on how routing improves confidentiality, integrity and availability.

**Table A3.2** How routing ensures confidentiality, integrity, and availability

|  | Confidentiality | Integrity | Availability |
| --- | --- | --- | --- |
| Routing. | Creating VRDs reduces the risk of unauthorized access to information. | Within the VRDs, information is maintained without modifications. | Each VRD is assigned specific throughput and QOS characteristics. |

Table A3.3 includes information related to the benefits of user segregation.

**Table A3.3**   How user segregation ensures confidentiality, integrity and availability

|                    | Confidentiality | Integrity | Availability |
|--------------------|-----------------|-----------|--------------|
| User segregation.  | STB and other equipment at the customer premises is secure from access by other subscribers. | Intruders are not able to cross-talk with other STB and manipulate their information. | Attackers are not able to spoof launch DOS attacks against other STB and customer equipment. |

Table A3.4 provides information on how QoS improves availability.

**Table A3.4**   How QoS ensures availability

|                     | Confidentiality | Integrity | Availability |
|---------------------|-----------------|-----------|--------------|
| Quality of service. | —               | —         | Attackers will not be able to use all the bandwidth and cause a DOS. |

## A3.4.4  User Segregation

With the imminent threat from computer viruses and worms, along with the clear risk of subscribers breaking into their neighbors' set top boxes, it is necessary to segregate users within their own networks. Subscriber segregation ensures that subscribers are not able to access each other's set top boxes, reducing the impact of computer worms and viruses. Cross-talk between DSL subscribers is blocked and all traffic is sent to the aggregation elements. Users are allowed to access only the head end for IPTV services.

## A3.4.5  Quality of Service

The DSLAM is capable of classifying traffic to ensure that QoS is maintained at all times. The DSLAM enforces rules to guarantee that critical services are always available. This provides additional protection against DOS attacks by subscribers trying to use all the bandwidth available for HTTP access to the head end. Sessions are controlled to ensure that there is bandwidth available for critical services.

Within the IPTV environment it is critical to ensure that subscribers are not able to flood the platform by sending too many requests to the head end. A minimum bandwidth must be guaranteed to ensure that subscribers are able to connect to the head end at all times to request services. Additionally, the broadcasted content sent by the head end requires minimum bandwidth levels to ensure customer satisfaction.

## A3.4.6 *Virtual Networks and Virtual Circuits*

When virtual private networks (VPNs and VLANs) are used to transmit valuable information using a public network, data is protected from unauthorized access.

Within the VPN, a group of authorized network elements is able to access the content. Virtual networks and virtual circuits can be used to segregate the IPTV traffic from all other traffic flowing through the DSLAM. With QoS and VLANs, the traffic can be controlled in a way that ensures that the IPTV VLAN has guaranteed bandwidth and service is maintained despite attempted DOS attacks.

Table A3.5 includes information on how VLANs improve security.

**Table A3.5**   How user virtual networks and virtual circuits ensure confidentiality, integrity and availability

|  | Confidentiality | Integrity | Availability |
|---|---|---|---|
| Virtual networks and virtual circuits. | Information is segmented which allows better protection of confidentiality of information. | There are clear restrictions on the type of traffic allowed, reducing the type of attacks that can affect the integrity of data and systems. | Traffic is segmented, reducing the possibilities of flooding. |

VLAN and/or VPNs can be used to segregate the traffic between specific circuits dedicated to services such as IPTV, VoIP, Internet access, control, etc. Within each circuit, different rules and security mechanisms can be deployed, certain traffic can be blocked and the amount of noise within the network can be dramatically reduced. For example, in the VLAN dedicated to IPTV, security mechanisms can be deployed to ensure that only HTTPS requests are sent between the STB and the head end.

Depending on the type of equipment available, this function can be deployed using layer-2 or layer-3 filtering – these are linked to other security requirements and capabilities of the equipment.

## A3.5  Conclusion

The DSLAM represents the first line of defense within an IPTV environment. It provides basic functions that ensure high levels of control over the type and amount of packages leaving the customer premise end-equipment. With a properly secured DSLAM, security experts are able to reduce the possibilities of fake packets on the network, thus increasing confidence in the communication. The security implemented at the DSLAM provides an initial layer that is reinforced by other levels of security higher in the path.

# Index